THE ECONOMICS OF THE OIL CRISIS

The Economics of the Oil Crisis

Edited by

T. M. RYBCZYNSKI

with a Foreword by

SIR FRANK MCFADZEAN

HM HOLMES & MEIER PUBLISHERS, INC.
New York

First published in the United States in 1976
by Holmes & Meier Publishers, Inc.
101 Fifth Avenue
New York, New York 10003

Library of Congress Cataloging in Publication Data

Main entry under title:

The Economics of the oil crisis.

Bibliography: p.
1. Petroleum industry and trade—Finance—Addresses, essays, lectures. 2. Petroleum products—Prices—Addresses, essays, lectures. 3. Power resources—Addresses, essays, lectures. I. Rybczynski, T. M.
HD9560.4.E24 338.2′3 75–34147
ISBN 0–8419–0235–6

Printed in Great Britain

Trade Policy Research Centre

The Trade Policy Research Centre in London was established in 1968 to promote independent analysis and public discussion of commercial and other international economic policy issues. It is a privately sponsored non-profit organisation and is essentially an entrepreneurial centre under the auspices of which a variety of activities are conducted. As such, the Centre provides a focal point for those in business, the universities and public affairs who are interested in international economic questions.

The Centre is managed by a Council which is headed by Sir Frank McFadzean, Chairman of the "Shell" Transport and

Trading Company. The members of the Council, set out above, represent a wide range of experience and expertise.

Having general terms of reference, the Centre does not represent any consensus of opinion. Intense international competition, technological advances in industry and agriculture and new and expanding markets, together with large-scale capital flows, are having profound and continuing effects on international production and trading patterns. With the increasing integration and interdependence of the world economy there is thus a growing necessity to increase public understanding of the problems now being posed and of the kind of solutions that will be required to overcome them.

The principal function of the Centre is the sponsorship of research programmes on policy problems of national and international importance. Specialists in universities and private firms are commissioned to carry out the research and the results are published and circulated in academic, business and government circles throughout the European Community and in other countries. Meetings and seminars are also organised from time to time.

Publications are presented as professionally competent studies worthy of public consideration. The interpretations and conclusions in them are those of their authors and do not purport to represent the views of the Council and others associated with the Centre.

The Centre, which is registered in the United Kingdom as an educational trust under the Charities Act 1960, and its research programmes are financed by foundation grants, corporate donations and membership subscriptions.

Contents

Foreword

Compared with the bilateralism and autarchical policies of the inter-war period, the period since the Second World War has witnessed a considerable advance towards a more open world economy. But nationalism remains a dominant force in international relationships and the institutions established to foster and encourage economic integration and interdependence are still relatively fragile. Problems and crises requiring global solutions are all too often the subject of cold calculation of national advantage. This was certainly true of the oil supply crisis which erupted in October 1973. It found the member governments of the Organisation for Economic Cooperation and Development (OECD) without an agreed plan for the allocation of reduced oil supplies.

This failure was not because the action of the Arab governments took the major consuming governments by surprise. Crises only give the impression of striking with the suddenness of lightning. The bulk of them in fact build up over a period of time. True, few foretold the precise date of the outbreak of the War of *Yom Kippur*, but the possibility of an oil shortage arising for one reason or another had been foreseen for an appreciable period. Some of the major oil companies had suggested that the OECD should at least consult together with a view to developing an emergency allocation system if circumstances required its introduction. Neither an unawareness of the problems, nor a lack of study of them, was the main cause of inaction and indecision in the OECD; it was an unwillingness to face up to the hard political choices which were involved.

There is no such thing as a perfect scheme for rationing supplies in time of shortage. Shortage involves sacrifice. The incidence of the latter can be varied; it cannot be avoided. It is a cause of much dissension and animosity. Should the rationing mechanism deal only with oil traded internationally or should local production also figure in the calculations? Should oil alone be considered or should the proposals also embrace the availability of other forms of primary energy such as natural gas in the Netherlands and coal and natural gas in the United Kingdom? Some countries invested in substantial reserve stocks of oil to meet an emergency, while others held little above the amount required for normal commercial purposes. How should varying stock levels be treated in a pro-rationing scheme? These, and many similar, points can be a fertile source of genuine disagreement in which the basic necessity of some rough and ready justice can easily become submerged. Every scheme has defects and if attention is focused entirely on these, paralysis of action will result. It was the fact that no agreement could be reached on the nature and distribution of the sacrifices that found the consumer governments unprepared to deal collectively with the crisis when it came.

This failure of the governments to reach agreement left a vacuum which was filled by the multinational oil enterprises. They proved to be the only organisations capable of taking a global view of the problems as opposed to "the partial and particular" national interests of each separate country. While observing the embargoes and other restrictions imposed by some of the Arab governments, the main international oil groups used their supply flexibility with non-embargoed crude oils to obtain as equitable a distribution as possible. Although all governments were informed of the cuts in supplies, there were only a few of them that did not, in one way or another, try to transfer their shortages to others; and there were only a few that did not claim either to be special cases or to have received assurances from one or other of the producing governments that they should receive their full requirements.

Politicians and friends of politicians descended in large numbers on the oil-exporting countries. Panic buying drove the price of the small quantities of oil owned by the host governments to over $20 a barrel. Adam Smith's view, expressed almost two centuries ago, of politicians as insidious and crafty animals "whose counsels are dictated by the momentary fluctuation of events" was amply justified in 1973. A situation which called for international statesmanship brought forth instead the disquieting face of chauvinism. One of the consequences was to underline the power that the oil-producers held, which in turn resulted in the very sharp increase in prices on 1 January 1974.

It is easy to envisage the economic difficulties and the acrimony that would have resulted if the major oil companies had succumbed to some of the pressures that were placed on them. It is more difficult to assess what the course of events would have been if the main Arab producers had carried out their threats to deepen their cut by 5 per cent every month until "the legal rights of the Palestinian people are restored". Considerable strains were building up and the pressure on the oil companies was increasing when normal production was restored.

The crisis left many legacies. Politicians trying to cash in on the discontent that followed the steep rise in prices – in terms of marker crude, Arabian light, producer government revenue increased by a factor of over five in a year – made many accusations against the conduct of the main multinational oil enterprises. These remain unsubstantiated; indeed, several inquiries in different countries have proved them to be groundless. In the United States a report of the Federal Energy Administration to Frank Church's Senate sub-committee stated that a more equitable distribution of reduced supplies during the crisis is difficult to imagine. Some of those who must share responsibility for the collapse of leadership are now quite vocal about the need to have a code of conduct for multinational enterprises. A case can indeed be made for this. But the more urgent need is to devise codes of conduct for governments which prevent the pursuit of national interests undermining broader international considerations.

The fact that the disruption was minimal has resulted in some questioning whether there was an oil crisis, at any rate of supply, as a result of the series of measures which some of the Arab members of the Organisation of Petroleum Exporting Countries (OPEC) inaugurated in October 1973. The facts are that in the world outside the Communist areas and North America, crude oil production in the month before the crisis was 36.2m barrels a day. By November it was reduced to a low of 32.3m barrels a day with the overhanging threat that the screw would be turned still further. When controls were eventually released, production increased, but it never came back to the September 1973 level. It peaked out in May 1974 at just below 36m barrels a day and fell rapidly thereafter. For the first six months of 1975 it oscillated around a figure of 29.5m barrels a day.

Oil consumption peaked in November 1973 at just over 37m barrels a day. Apart from seasonal variations, consumption continued to fall until by mid-1975 it was around 29m barrels a day, or close to the level prevailing three years previously. The fall in demand was the result of several factors – a worldwide economic downturn, conservation measures, resistance to the higher prices and a succession of unusually mild winters. The industry had invested to meet a much higher level of demand and currently there are over 10m barrels a day of crude oil shut in; nearly two years after the crisis broke there are 419 tankers totalling some 34m tons deadweight laid up, with an additional 3¼m tons in the Persian Gulf awaiting cargo, while refineries outside the United States were functioning at substantially less than 70 per cent of their distillation throughput capacity.

Owing to various time lags as a result of stocks and the long supply lines, the unprecedented increase in crude oil prices took some time to affect the balance of payments of the importing countries, but on the whole few have doubted that these have critical implications. There remains a great deal of confusion however as to just what these implications are, what their time scale is and how they differ in their impact on various countries and on the international economic order as a whole. The essays contained in this volume collectively describe and analyse the range of problems, while the Bellagio Memorandum embodies a concern for the maintenance of multilateral cooperation, now threatened from many directions.

How far the OPEC-enforced increases in producing governments' revenues have contributed to the present worldwide recession is difficult to assess. So far as inflation is concerned, Harry Johnson, with his usual perspicacity, points out in his essay that "rises in money prices are themselves deflationary, not inflationary; they are a way of eliminating excess demand for goods and excess money supply . . . the inflationary process consists in continuous recreation of inflationary conditions requiring further price rises to correct them . . . to check inflation it is necessary to break inflationary expectations." Nothing has done more to create and recreate these pernicious conditions than the virtual collapse of financial discipline over large areas of central and local government expenditure. The continued politicisation of power has been one of the main causes of the degeneration into near financial anarchy. Again Adam Smith saw the nature of the problem all too clearly. "It is impertinence and presumption on the part of governments to pretend to watch over the economy of private people," he wrote, "when they are themselves, always and without any exception, the greatest spendthrifts in the society. Let them look well after their own expense and they may safely trust private people with theirs. If their own extravagance does not ruin the State, that of their subjects never will."

In this volume George Ray covers extensively the impact of the oil crisis on the energy situation in Western Europe, while T. M. Rybczynski deals with the capital requirements for developing alternative sources of energy. The general theme of the essays is the need for government to seek some form of accord to meet the world's interests as a whole.

Almost two years after the crisis no international mechanism has been introduced to ensure that the OPEC "surpluses" will be recycled back to the consumer countries *pro rata* to their requirements. To prevent the petro-dollars disrupting international financial and trading mechanisms is still the most immediate problem. For 1974 these sums have been put at some \$60,000–70,000m. They are bound to grow cumulatively through additional sales of oil, but also as a result of the interest payments which the balances earn. It is unlikely that the cumulative total will by 1980 fall below some \$200,000m, while \$650,000m – the figure they would reach if 1973 rates of exports from OPEC sources and 1974 rates of revenue per barrel were to continue unchanged – probably constitutes a top figure. A mid point of some \$450,000m by 1980 represents a fair conjecture.

These funds are of course the obverse of the current balance-of-payments deficits of the oil-importing countries. Absolutely, they are made up chiefly by the industrialised countries of Japan, Western Europe and the United States, but in relative terms the developing countries have been hard hit – India particularly so. On an unrestricted basis, oil imports would have absorbed half of India's export earnings in 1974.

Deficits of this order offer a standing threat to the relatively liberal order of financial and trading arrangements which have been a condition of the unprecedented economic prosperity enjoyed since the Second World War. In the absence of international agreement as to how the oil surpluses are to be recycled, governments could turn to import restrictions, export sudsidies and similar beggar-my-neighbour expedients to achieve balance in their international current accounts. To do so would be self-defeating for the world as a whole and, since an "improvement" for one importing country could only be obtained at the cost of a deterioration for another, incentives to retaliatory strife would progressively multiply. Moreover, such an outcome is entirely unnecessary, for a variety of workable recycling schemes have already been put forward. They vary in their merits and demerits. None is perfect, but perfection is not called for, just a consensus in favour of one or more of the proposals which are in any case not mutally exclusive.

What is required, therefore, is *au fond* an act of political will by the world's principal trading nations. If it is not forthcoming, disruption could certainly occur and the steady growth of international trade, with all that this implies for the domestic objectives of rich and poor countries alike, could come to seem a purely historical phenomenon of the "pre-crisis" decades.

FRANK McFADZEAN

London
July 1975

Preface

This is the first volume of papers to be published in the Trade Policy Research Centre's programme of studies on the Reform of the International Trading System. The programme has been conducted with the help of generous grants made available by the Leverhulme Trust Fund, in London, and the Ford Foundation, in New York. Its conclusions are being drawn together in a report that is being prepared by a study group under the chairmanship of Sir Frank McFadzean, who contributed the above Foreword. The studies, though, have been supervised by David Robertson, Reader in International Economics at the University of Reading.

Several of those involved in the programme enjoyed the hospitality of the Rockefeller Foundation at the Villa Serbelloni, Bellagio, by Lake Como in Italy, for a preliminary meeting at which was drafted a statement on the international economic situation in the light of the dramatic increase in crude oil prices at the end of 1973. The statement was subsequently finalised as the Bellagio Memorandum and published in June of the same year. The text is reproduced on pages xxiii–xxxv of the present volume.

Not all the papers in the volume were a part of the original programme. But when the oil crisis was precipitated it was clear that account would have to be taken of its economic implications. Some of the papers presented at meetings of the Centre have been published in other places and appropriate acknowledgements are made in the notes which follow them here. Of those published earlier by the Centre, the chapter by W. M. Corden and Peter Oppenheimer first appeared as Staff Paper No. 6, *Implications of the Rise in Oil Prices*; the two chapters by George Ray as Thames Essay No. 3, *Western Europe and the Energy Crisis*.

Editing a volume of essays is not an easy task. In working on the papers, improving their presentation, and up-dating factual details, Mr Rybczynski was assisted by Janet Strachan, Anne

Souchon and Cedric Watts, while Margaret Bacciorelli prepared the final typescript.

It should be emphasised that the views expressed in the papers do not necessarily reflect the views of the Council or of others associated with the Trade Policy Research Centre. For the Centre, having general terms of reference, does not represent any consensus of opinion. Its purpose is to promote independent research and public discussion of international economic policy issues.

HUGH CORBET
Director
Trade Policy Research Centre

London
Spring 1975

Biographical Notes

W. M. CORDEN, Nuffield Reader in International Economics at the University of Oxford, has made numerous contributions to the development of international trade theory, particularly on the concept of effective protection. Dr Corden was previously a Professorial Fellow in the Research School of Pacific Studies at the Australian National University. He is the author of *The Theory of Protection* (1971) and *Trade Policy and Economic Welfare* (1974), both major works, while shorter contributions to public discussion have included *Monetary Integration* (1972) and, with Ian Little and Maurice Scott, *Import Controls versus Devaluation* (1975).

HARRY G. JOHNSON has been Charles F. Grey Professor of Economics at the University of Chicago since 1974, having occupied a chair there since 1959, during which period he was also, from 1966 to 1974 Professor of Economics at the London School of Economics and Political Science. Since its inception in 1968, Professor Johnson has been Vice Chairman, and Director of Studies, of the Trade Policy Research Centre in London; and he is active in other bodies concerned with the promotion of independent research and public discussion of economic policy problems. From 1956 to 1959 he was Professor of Economic Theory at the University of Manchester, having earlier been a Fellow of King's College, University of Cambridge (1950–56).

Among Professor Johnson's books are: *The World Economy at the Crossroads* (1965); *Economic Policies Toward Less Developed Countries* (1967); *Essays in Monetary Economics* (1967); *Comparative Cost and Commercial Policy Theory for a Developing World Economy* (1968); *Aspects of the Theory of Tariffs* (1971); *Inflation and the Monetarist Controversy* (1972); *Further*

Essays in Monetary Economics (1973); *On Economics and Society* (1975); and *Technology and Economic Interdependence* (1975).

ALASDAIR MACBEAN has been, since 1968, Professor of Economics at the University of Lancaster, having previously been Economic Adviser to the British Ministry of Overseas Development. After obtaining his doctorate at Harvard University, Professor MacBean was, from 1962 to 1964, an Economic Adviser to the International Bank for Reconstruction and Development. He takes a special interest in problems of trade and development, being the author of *Export Stability and Economic Development* (1966) and co-author, with V. N. Balasubramanyam, of a major review of policies towards developing countries, *Meeting the Third World Challenge*, to be published in 1976.

SIR FRANK MCFADZEAN, who has contributed the Foreword to this book, is Chairman of the Shell Transport and Trading Co. Ltd, London, and a Managing Director of the Royal Dutch–Shell Group of Companies. He has been Chairman of the Trade Policy Research Centre since 1971, and is Visiting Professor of Economics at the University of Strathclyde, where he is also Chairman of the Steering Board of the Strathclyde Division of the Scottish Business School. Before joining Shell in 1952, Sir Frank worked in the British Treasury and the Board of Trade, and was area chief in Malaya of the United Kingdom's Colonial Development Corporation. On his retirement from Shell in 1976, Sir Frank is to become Chairman of British Airways.

PETER OPPENHEIMER, a specialist on international monetary problems, has been a Student (i.e. Fellow) of Christ Church at the University of Oxford since 1967; and before that he was a Research Fellow at Nuffield College, also at Oxford. From 1961 to 1964, Mr Oppenheimer was on the staff of the Bank for International Settlements in Basle. He has contributed to numerous professional journals and other publications and also broadcasts on radio and television.

GEORGE F. RAY is a Senior Research Fellow at the National Institute of Economic and Social Research in London where, since 1968, he has been on the editorial board of the *National Institute Economic Review*, which reports quarterly on the state of the British economy. Born and educated in Budapest, Hungary, Mr Ray joined the National Institute in 1957 and takes a special interest in energy matters. Mr Ray is the author of *The Competitiveness of British Industrial Products* (1966) and co-author of *The Diffusion of New Industrial Processes: an International Study* (1974), besides a number of other studies.

T. M. RYBCZYNSKI, Economic Adviser at Lazard Bros & Co. Ltd, London, is Chairman of the Society of Business Economists and a Visiting Professor of Economics at the City University of London. Before graduating, he contributed to the subject of economics "the Rybczynski theorem", one of the fundamental principles of international trade theory. For a time Mr Rybczynski was a part-time lecturer in Economics at the London School of Economics and Political Science.

JAN TUMLIR has been Director of Research at the GATT Secretariat in Geneva since 1967. After reading law at Charles University in Prague, where he was one of the publicists in the student resistance to the Communist *coup d'état* in 1948, he escaped from Czechoslovakia and worked for a time in Germany before going to the United States, where he read economics at Yale University. On obtaining his doctorate, he taught economics at Yale, returning to Europe in 1964 to take up an appointment at the GATT Secretariat.

Bellagio Memorandum

In the wake of the dramatic increase in oil prices towards the end of 1973, the memorandum below was prepared, for the purposes of public discussion, by a group of fifteen European businessmen and commentators on international economic affairs.

As guests of the Rockefeller Foundation, the group prepared the first draft at the Villa Serbelloni, Bellagio, overlooking Lake Como in Italy, meeting there for three days in April 1974 under the chairmanship of Sir Frank McFadzean, the chairman of the Shell Transport and Trading Company. The draft was finalised as the Bellagio Memorandum and issued in June of that year under the auspices of the Trade Policy Research Centre. The signatories were.

Sir Frank McFadzean: Chairman, Shell Transport and Trading Co. Ltd, London, and Chairman, Trade Policy Research Centre, London.

Sir Alec Cairncross: Master, St Peter's College, University of Oxford; formerly Head, British Government Economic Service.

Mr Hugh Corbet: Director, Trade Policy Research Centre, London.

Professor Gerard Curzon: Professor of International Economics, Institut Universitaire de Hautes Etudes Internationales, Geneva.

Dr Victoria Curzon: Research Fellow, Institut Universitaire d'Etudes Europeennes, Geneva.

Dr Emanuele Dubini: Deputy Chairman, Pirelli S.p.A., Milan; formerly Deputy Chairman, Confederazione Generale Industria Italiana, Rome.

Professor Sir Ronald Edwards: Chairman, Beecham Group Ltd, London, and Professor of Industrial Organisation, London School of Economics and Political Science.

PROFESSOR HANS KRAMER: Professor of International Organisation, Institut für Weltwirtschaft, University of Kiel.

DR PAUL KREBS: Director and General Manager, Deutsche Bank A.G., Frankfurt.

PROFESSOR THEO PEETERS: Professor of International Economics, Centrum voor Economische Studiën, Catholic University of Leuven.

MR ALAN PETERS: Economist, Shell International Petroleum Co. Ltd, London.

MR JEAN ROYER: Adviser on Trade Policy, International Chamber of Commerce, Paris; formerly Deputy Secretary-General, GATT Secretariat, Geneva.

MR T. M. RYBCZYNSKI: Economic Adviser, Lazard Bros and Co. Ltd, London; and Chairman, Society of Business Economists, London.

DR VITTORIO SALLIER DA LA TOUR: Economic Adviser, FIAT S.p.A., Turin.

DR H. E. C. VAN BEURDEN: Director (External Relations), N. V. Philips, Eindhoven.

In the text of the memorandum as reproduced here, the original notes are numbered, whereas notes made for the purpose of updating the discussion are lettered. The text follows:

REFORM OF THE INTERNATIONAL COMMERCIAL SYSTEM

Since the autumn of 1973 there has been a quantum increase in the degree of uncertainty surrounding the conduct and planning of company, industry and national economic affairs.

Major Sources of Uncertainty

Where previously there was a vague though general premonition of an energy and particularly a petroleum shortage, there is today acute and generalised uncertainty about the effects of the dramatic increase in crude oil prices, the economic implications of which are very far reaching. The net liquid assets accruing to the oil-producing countries could increase by up to \$70,000m per annum (as compared with around \$100,000m in the Eurodollar market at the present time), the actual size of the problem in coming years depending on the quantity of oil demanded and supplied and on the price charged. If prices of crude oil remain around their current level, and if oil-importing countries are prepared to carry the debt involved, there could conceivably accumulate by 1980 an "oil-debt overhang" of $500,000m or so.

It is generally agreed that for many years the oil-producing countries will not be in a position to absorb real transfers of goods as a counterpart to this debt and, as a result, unprecedented financial flows are certain to be generated. These large flows, the full force of which has yet to be felt, could have a distorting and disrupting impact on exchange-rate relationships. Some countries may be tempted, or may even be obliged, to achieve a balance on their current accounts by means of competitive devaluations, import restrictions or domestic deflation which in the circumstances – combined, that is, with other factors – could lead to the disintegration of the international economic order.[1] It is therefore not so much the size of the transfer problem that is the cause of concern. What matters more is how the problem is managed by the industrialised countries.

While private financial markets can handle – without putting pressure on the system – a considerable part of the financial flows that are anticipated, intergovernmental action is vital in two areas if the threat to the international economic order is to be averted:

(a) Through existing multilateral institutions, governments should agree on means to cope with the recycling, and/or funding, problems which private financial markets would find difficult to manage.[a]

(b) The temporary "standstill" on trade measures that might otherwise be introduced to offset the effects on balances of payments of the increased price of oil, as agreed within the framework of the Organisation for Economic Cooperation and Development (OECD), should be maintained until more enduring provisions are negotiated.[2]

These actions should allay fears for the international commercial and monetary systems. In the longer term, however, the oil-importing countries will only be able to work off the oil-debt overhang by using some of the credit provided to them by the oil-producing countries to increase productive capacity; it will not be sufficient merely to maintain or increase present levels of consumption.[3,b]

A second major source of uncertainty is the likelihood that the problem of inflation will be magnified still more. Moreover there is now growing uncertainty arising out of the possibility of a general recession. Governments must cooperate in the management of their respective economies if they are to avoid the instability in commercial and monetary relationships that could result from a recession or from accelerating inflation or from both.

Third, and most ominous perhaps, is the political uncertainty emanating from these developments: the uncertainty over the future (i) of the alliances on which international economic cooperation since World War II has been built and (ii) of the intergovernmental organisations through which it has been pursued. If the governments of open societies do not want their system of collective security to disintegrate they will have to make a more conscious effort to resolve economic issues between them on a cooperative basis. And they should do so in the existing multilateral institutions to which countries large and small look for their interests to be protected.

All these uncertainties affect firms and industries in every country and to much the same extent; furthermore, they feed upon each other. In their present intensity they are straining the pattern of economic, social and political life as it has been organised in the postwar period, both within and among countries. The consequences of such uncertainties continuing much longer are truly immeasurable. And no government is in a position to settle them on its own.

The first act required, then, to contain the pervasive economic problems of our time was a convincing expression of political will on the part of governments to solve common problems in a multilateral context. Two fora were accordingly established to broach the most important issues:

(a) the Committee of Twenty and its successor, the executive committee of the board of governors of the International Monetary Fund (IMF), for a negotiated reform of the international monetary system; and

(b) the Tokyo Round of multilateral negotiations under the General Agreement on Tariffs and Trade (GATT), for a reform of the international commercial system.

In addition, the OECD is becoming a useful forum for preparing, or discussing, proposals on a number of issues of primary interest to developed countries before they are taken up in wider institutional settings.

Development of the Crisis

Confidence in the continued stability and expansion of the world economy was gravely shaken by the rapid succession of economic crises which began in the late 1960s and culminated in the breakdown of the international monetary system in 1971. The oil crisis at the end of 1973 worsened the situation. From the outset it was recognised that these crises demonstrated weaknesses which, if not corrected, could lead to increasing economic uncertainty and undermine the international economic order – painstakingly restored in the late 1940s and 1950s, following the disorders of the early 1940s and 1930s.

These weaknesses have been due in part to the inability of intergovernmental institutions to respond to economic changes. Governments have not been able to keep pace with the growing interdependence of national economies which resulted from the postwar integration of the world economy. Economic interdependence, and the trend for governments to assume ever wider responsibilities in domestic affairs, has heightened the possibility of economic measures in one country seriously disrupting economic conditions in other countries.

Recognising the dangers inherent in the erosion of the framework of multilateral cooperation, the major trading countries agreed in February 1972 to embark on a reform of the international economic order. What is needed – more than two years later[c] – is for this endeavour to be afforded urgent and higher priority in order to generate the necessary momentum to carry negotiations through to a successful conclusion.

On the monetary front, discussions held in the context of the Committee of Twenty have provided the principal elements for a reformed international monetary system. These discussions have also succeeded in establishing an informal basis for continuous cooperation between governments on monetary questions. The executive committee of the IMF has still to give these cooperative arrangements more concrete form. Above all, though, there is an overriding need for agreed rules on the management of floating rates of exchange.[d] Common rules have to be set for intervention by national authorities in currency markets in order to minimise the danger of disorderly exchange-rate movements which could result either from conflicting intervention policies or from the absence of intervention.

On the commercial front, the ministerial meeting in Tokyo in September 1973 agreed a sound basis for a seventh round of GATT negotiations, aimed at improving arrangements for regulating trade policies and continuing the removal of tariffs and the modification of non-tariff distortions of both industrial and agricultural trade. While the preparatory work has begun in Geneva there has not been enough effort in the European Community to prepare for these negotiations. Actual negotiations cannot begin until the Administration of the United States has secured Congressional authority.[e] But this need not – and must not – delay the formulation of detailed, though flexible, negotiating directives for the Community's representatives. The "overall approach" approved by the Council of Ministers falls far short of what is required for specific discussion.[4,f]

In particular, the governments of the European Community should prepare themselves, *as well as business and public opinion*, to

(a) improve, through institutional and procedural changes, the management of multilateral commercial relations in order to prevent any retrogression from the levels of trade liberalisation already achieved, and

(b) enlarge the degree of trade liberalisation by reducing further the levels of protection to producers in order to maintain and increase the responsiveness of national economies to international market forces.

Both objectives should include a deliberate effort to safeguard international trade flows from arbitrary government intervention. Provisions should be developed for ensuring international compatibility between the social, regional, environmental, resource-development and other policies which governments pursue. When the price mechanism is set aside, misallocations of resources are liable to ensue, resulting in loss of international competitiveness in the industries involved. In the absence of multilateral coordination, policies of government intervention can easily conflict with one another, thereby impairing the degree of freedom of trade negotiated over the last quarter of a century.

Relevance of Trade Liberalisation

What is being reported from Geneva, and from national capitals, about the "position-building" of governments for the negotiations seems to reflect a mercantilist "you gain/we lose" attitude. To realise how utterly unrealistic this attitude has become one only has to contemplate the consequences of failing to arrive at a substantial negotiated result. The real business of the Tokyo Round negotiations can begin only when such attitudes are abandoned.

It has recently become fashionable to say that emphasis on trade liberalisation reflects a concern of a bygone era in which the main problem was excess capacity and unemployment – and hence access to foreign markets for one's products – and that in the present context, dominated by inflation and specific supply scarcities, trade liberalisation is irrelevant and different solutions have to be sought. But the traditional balance of arguments for opening up national markets by lowering obstacles to trade remains unchanged. The benefits of international specialisation in production still exist. Total income can be maximised if each participant in world trade is allowed to maximise what he can produce most efficiently.

Liberalisation measures alone, however, are not enough to ensure the stability of existing trade flows. While there is a need for the further liberalisation of world trade, it must be acknowledged that in the process of liberalisation situations may arise in which individual countries may have to limit temporarily their imports of individual products. Indeed, the progress of liberalisation in general may be contingent on temporary limitations in specific cases, as well as on structural adjustment. If this principle is recognised it provides a basis on which rules can be formulated ensuring adequate security and equity for both exporting and import-competing industries in critical situations.

The new problem of shortages of many primary commodities and semi-manufactures makes further trade liberalisation more essential than ever. Many of these shortages have their origin in under-investment in the sectors concerned. Investment required for the production of most of these basic materials will not be undertaken without an assurance that the product can find a market. The key to secure supplies lies in assurance of access to markets. There is little doubt that an accelerated liberalisation of world trade would mobilise investment towards the most efficient locations of production and, in the process, help to eliminate the existing supply bottlenecks, restore the regular distribution of intermediate goods and thus constitute a most effective weapon against inflation.

Recent developments in the Organisation of Petroleum Exporting Countries (OPEC) have given rise to concern about assurance of "access to supplies" which is a counterpart to the more familiar concern about assurance of "access to markets". Most expressions of this new concern seem exaggerated. None the less, the possibility of efforts to form raw material cartels cannot be excluded; and even an ultimately unsuccessful effort could cause substantial disruption in and among globally interdependent economies. The GATT contains a number of rules which could give some assurance against disruptions of this kind if a will was demonstrated to use them. They should be elaborated and developed further and, if necessary, supplemented by additional rules.

Substantial Elimination of Tariffs

There is a danger of too much increasingly precious time being devoted to bargaining over tariffs. It appears that with respect to tariffs governments have of late created a new mythology. They have developed a finesse in tariff negotiations which is wholly illusory and completely out of proportion with what is foreseeable as a result of more or less general tariff reductions. The need for a broad give-and-take in trade negotiations is understood. But the dogged product-by-product bargaining of the Kennedy Round negotiations had about it an air of unreality. It seemed to be a self-indulgent game played by bargainers who had no way of assessing the results of specific concessions. How much less real would similar bargaining be today in an economic environment of floating currencies and annual rates of inflation perhaps double the average level of protection now provided by national tariffs?

At their present level, the tariffs of the major industrial countries provide significant effective protection mainly to industries producing basic semi-manufactures such as steel and non-ferrous metals, pulp and paper, wood panels, leather and processed food. In the majority of these cases the processing industries of the most industrialised countries are based on imported raw materials. The development of the world economy has tended to increase the bargaining power of raw material suppliers who want to establish processing industries of their own and so reap the value added in the processing of their primary resources. In all other areas of trade, tariffs can be phased out over a period with much less difficulty of consequent adjustment.

To say that the economic effect of tariffs – that is, the protection they afford – has greatly diminished does not imply that the further reduction and ultimate elimination of those that are outstanding has become less important or less urgent. The political cost of the continued existence of tariffs, particularly of the unavoidable pressures to grant preferences under them to certain countries, remains high; indeed, compared with their real effects, it is excessively high.

Since in one area the major trading powers will have to, and in most of the other areas they painlessly could, eliminate tariffs over a transitional period, governments should now seek simpler and less time-consuming methods of negotiating on tariffs than traditional bargaining for reciprocal concessions. Approaching the tariff negotiation on the understanding that the principle of reciprocity is to be interpreted in a broad sense, rather than on an item-by-item basis, would certainly facilitate the process. A declaration of intent by the major trading powers to proceed, perhaps in two stages, towards the elimination of tariffs on trade among developed countries by progressive, linear and automatic reductions according to an agreed timetable would be even more effective in this respect.[g]

Non-tariff Interference with Trade

Like tariffs, many non-tariff interventions are designed to protect or favour domestic producers *vis-à-vis* foreign suppliers – at the expense of domestic consumers and taxpayers. Others, though, are aimed at achieving a variety of social objectives, but have unintended trade-distorting side-effects. Still others are temporary measures used at industry or national level to enable a short or medium-term problem to be overcome.

The objective of the Tokyo Round discussions should be to remove or modify as many as possible of these measures which range from customs valuation procedures, import quotas, "voluntary" export restraints, industrial standards and government subsidies to public procurement policies. With many types of non-tariff intervention, however, a gradualist approach would be most feasible. This would involve agreement on codes of conduct followed by a process of more or less continuous consultation and negotiation on their implementation. Provision would need to be made for multilateral surveillance and complaints procedures.

In this way it should be possible in due course to ensure that the level of protection afforded by non-tariff interventions is the minimum necessary to achieve legitimate social objectives. Two guiding principles should therefore be adopted:

(a) First, intervention in the market should be restricted to those occasions when the behaviour of the market would produce major distortions of, or deviations from, desirable social objectives.

(b) Second, in formulating their policies, governments should avoid measures which have the effect of transferring to foreigners a large part of the costs involved in achieving domestic objectives.

Such should be the guiding principles, too, of endeavours in the European Community, and in the European Free Trade Association, to coordinate non-tariff interventions by governments.

Safeguards on Imports and Exports

An important category of non-tariff intervention, namely quantitative import restrictions, concerns the avoidance of disruptive market conditions in a particular industry that result from sharp increases in imports of competing products. The growth of trade has greatly accelerated in the postwar period and, with the progress of industrialisation in developing countries, shifts in comparative advantage are increasingly rapid and frequent. In this situation a "safeguard" provision in the GATT that can offer security and equity to both exporting and importing countries would have to be an instrument of change and adjustment. It would fulfil this function if it combined the right to temporary protection in specified circumstances with an obligation on the part of the protecting government, or of the industry seeking protection, to effect the structural adjustments indicated by the change in the trade pattern. Such safeguard measures should be taken within internationally established guidelines, including domestic efforts to facilitate adjustment, and they should be subject to multilateral surveillance. A safeguard provision in this form would also be the most effective instrument for coping with non-tariff restrictions inherited from the past.

Similarly, a sudden decline in available supplies of a critical commodity, either at home or abroad, may from time to time make it necessary for governments to restrain exports. Such safeguard actions on exports also (i) should be taken only for temporary periods to ease the process of adjustment, (ii) should recognise the interests of both the net exporting and net importing countries, and (iii) should be subject to continuing multilateral surveillance.

Case of Agricultural Trade

Non-tariff interventions are employed in the stabilisation of market conditions for certain commodities in the field of agricultural trade. Negotiations in this area should focus on the different policy approaches taken by governments to stabilise prices and incomes, and the implications which these policy choices have on market conditions in other countries and the possibilities for international trade. The objective of such negotiations should be to define a set of commitments which would accord equal consideration to the interests of net exporters and net importers of agricultural products and which would provide expanded opportunities for agricultural trade.

Interrelationship of Trade and Monetary Measures

The balance-of-payments difficulties expected by many oil-importing countries have created an urgent need for a revision of the rules governing the application of trade restrictions for balance-of-payments reasons. There is a need to revise both the GATT rules and the IMF provisions applicable to the use of trade measures for balance-of-payments purposes. Such rules should permit the use of tariff surcharges as well as quotas, although they should discourage their use except where that serves as a useful complement to other balance-of-payments adjustment measures, and they should be used only for temporary periods.

The strengthening of the intergovernmental arrangements to cope with the changing requirements of international economic relations should cover four broad areas: (i) the operation of the international payments system under the aegis of the IMF as a provider of a world *numéraire*; (ii) the application of trade and exchange measures to adjust to balance-of-payments difficulties for which the IMF and the GATT would be jointly and equally responsible; (iii) the gradual development of a long-term commercial policy in conjunction with the adjustment to be made in the industrial structure of the national and international economies; and (iv) an agreed code of conduct for international investment, including an adequate degree of security for providers of funds.

All these tasks are urgent and will have to be carried out

virtually at the same time within the next three to four years. They require of governments a readiness to limit the implications for other countries of their domestic adjustment measures and to consult more frequently on policy problems at appropriately responsible political levels. A significant step forward would be made if the major countries were prepared to agree rapidly on a declaration of their readiness to accept the implications of this new form of international cooperation. If the reform of institutional arrangements is conducted in such a spirit there is hope of economic growth proceeding in an orderly fashion to the benefit of all trading nations.

NOTES AND REFERENCES

1. By the "international economic order" is meant the institutions established in the late 1940s to manage the world economy, namely the International Monetary Fund (IMF), the International Bank for Reconstruction and Development (IBRD), the General Agreement on Tariffs and Trade (GATT) and, in the 1960s, the Organisation for Economic Cooperation and Development (OECD).
2. The "standstill" was first proposed by the United States at the meeting of the Committee of Twenty in Rome, January 1974, and finally agreed upon at the ministerial meeting of the OECD in Paris at the end of May. But it was only agreed for one year. (At the 1975 ministerial meeting, however, it was renewed for another year.)
3. One solution would be to channel all transactions with the oil-producing countries through the Bank for International Settlements (BIS) – perhaps employing a weighted average of currencies. The oil-producing countries would build up interest-bearing balances with the BIS upon which they could draw for their current import needs. Only at that moment, however, would oil-importing countries be called upon to make a transfer of currency to the BIS on a pro-rata-of-oil-imports basis.

   In this way the oil-producing countries would maintain the value of their assets (and earn a return on them) and the oil-importing countries would retain control over their monetary and, in particular, their exchange-rate policies. In this way, too, it would be possible to contain the disruptive monetary and fiscal effects of the flows being generated and so prevent the emergence of the overhang. The holders of the debt – the oil-producing countries – would have to be satisfied though that, in the ultimate, they would receive the real counterpart, in goods and services, of that debt.

   Such a scheme would necessarily involve the wholehearted participation equally of the oil-producing and the oil-consuming countries in a common effort of mutual self-interest.
4. *Overall Approach to the Coming Multilateral Negotiations in GATT*, Document 1/135 e/73 (COMMER 42), Commission of the European Community, Brussels. The document was agreed by the Council of Ministers on 26 June

1973. It should be noted that this final version differs in a number of important respects from the version submitted to the Council by the Commission on 4 April of that year.

5. For a fuller discussion of this and other proposals for maintaining the liberalisation of international trade, see Frank McFadzean *et al.*, *Towards an Open World Economy*, Report of an Advisory Group (London: Macmillan, for the Trade Policy Research Centre, 1972).

a. Proposals were advanced by Johannes Witteveen, as managing director of the IMF, at the January 1974 meeting in Rome of the Committee of Twenty, which were elaborated upon by Denis Healey, as Britain's Chancellor of the Exchequer, at the annual meeting of the IMF later that year. But governments have been slow to agree and implement arrangements for recycling, and/or funding, petro-dollars.

b. The distinction between borrowing for investment and borrowing for consumption was emphasised, in the British context, in W. M. Corden and Peter Oppenheimer, *Basic Implications of the Rise in Oil Prices*, Staff Paper no. 6 (London: Trade Policy Research Centre, 1974), republished in the present volume as Chapter 3.

c. As the present volume goes to press "more than *three* years later" the need remains as great as ever.

d. In the end a set of rules was formulated in the Committee of Twenty and agreed at the 1974 meeting of the IMF.

e. The Congress of the United States afforded the administration an authority to negotiate when it passed the Trade Act of 1974 in December 1974. The legislation was signed by President Ford on 5 January 1975.

f. The negotiating directives for the Commission of the European Community were agreed by the Council of Ministers just a few days before the Tokyo Round of GATT negotiations got down to business in February 1975 – having been formally opened in September 1973.

g. Neither the United States nor the European Community is authorised to negotiate the substantial elimination of industrial tariffs in the Tokyo Round negotiations, although Japan is authorised to go that far, but a formula for linear tariff reductions could be worked out on the basis of substantial elimination being achieved in the next GATT round.

List of Abbreviations

c.i.f.	prices including cost, insurance and freight
ECSC	European Coal and Steel Community
EFTA	European Free Trade Association
f.o.b.	free-on-board prices
GATT	General Agreement on Tariffs and Trade
GDP	Gross Domestic Product
GNP	Gross National Product
GSP	Generalised System of Preferences
IBRD	International Bank for Reconstruction and Development
IEA	International Energy Agency of the OECD
IMF	International Monetary Fund
MFN	most-favoured-nation clause, expressing the principle of non-discrimination; also relates to non-discriminatory rates of duty
OECD	Organisation for Economic Cooperation and Development
OEEC	Organisation for European Economic Cooperation (replaced by the OECD)
OPEC	Organisation of Petroleum Exporting Countries
SDRs	Special Drawing Rights on the IMF
SITC	Standard International Trade Classification
t.o.e.	tons of oil equivalent
UNCTAD	United Nations Conference on Trade and Development
World Bank	International Bank for Reconstruction and Development

CHAPTER 1

Historical Background to the World Energy Crisis

T. M. RYBCZYNSKI and GEORGE F. RAY

Coming on top of chronic inflation and serious commodity shortages, the dramatic increase in crude oil prices towards the end of 1973 shook the world economy to its very foundations, precipitating the energy crisis which had long been anticipated – if not for another decade or so. How to cope with the economic implications of the crisis became, almost overnight, the major preoccupation of governments all round the world. In the industrialised countries, virtually everybody – in every walk of life – has been affected by higher prices for all types of petroleum products, never mind the increased demand for substitutes and the further twist to the inflationary spiral. But it is the likely consequences of huge revenues accruing to the oil-producing countries that is perhaps the principal cause of concern.

What, then, is meant by the "energy crisis"? As a "catch all" phrase it encompasses several aspects of the situation. It embraces, for a start, the long-term as well as the short-term aspect of the supply of non-renewable sources of energy. It embraces, too, the sudden change in relative prices of various types of energy, besides the shift in those prices *vis-à-vis* all other prices. Also embraced by the phrase is the impact of these price changes on the earnings and financial reserves of both oil-producing and oil-consuming countries. And the implications of this last, for the international system of trade and payments, must therefore be implicit in the phrase.

PRECIPITATION OF THE CRISIS

It was in the late 1960s[1] that forecasts of an imbalance between supply and demand for oil in the 1980s, assuming no radical change in prices, began to influence policy formation in the developed economies, especially in the United States. By the early 1970s a spectacular increase in American demand for imported oil had created a sellers' market. But calculations about the future were upset by the unilateral action of the Arab oil-producing countries in October 1973 when, with the outbreak of the War of *Yom Kippur*, they announced an immediate cut in their production of crude oil and an increase in its price.[2]

The quantitative restrictions on oil supplies were firmly tied to political objectives. Countries "friendly" towards Israel, chiefly the United States and the Netherlands, were formally boycotted while those "friendly" to the Arab cause were promised preferential treatment. The rest of the world was to receive reduced supplies which were to be tightened every month until a settlement with Israel – on Arab conditions – was reached.

No political conditions were attached to the higher prices that were demanded. The "posted prices" of crude oil – which had served as the basis for calculating the "take" of host governments – was increased by 70 per cent, with special premia for oil of low sulphur content and also, in the case of Libya, for proximity to Western Europe. For the countries of Western Europe as a whole, the rise worked out at about 80 per cent of the September 1973 price, the level that had been reached following the increases negotiated under the Tehran, Tripoli and Geneva agreements between the international oil industry and the governments of the Organisation of Petroleum Exporting Countries (OPEC).

On 16 December 1973 the new price was doubled again. Thus, at the beginning of 1974, importing countries were faced with crude oil prices that were three and a half to four times what they had been three months earlier. Taking Arabian light crude as a representative marker, the revenue per barrel for producing governments rose from $0.91 in 1970 to $1.27 under the Tehran Agreement of 1971, to about $3.30 in October 1973 and to around $8.00 in January 1974.

The strength of the OPEC countries waxed even greater over the ensuing year. In a variety of ways they were able to increase their total "take" from each barrel of oil. Certain "posted" and other prices were raised. Royalty payments were increased and a higher "participation" achieved. Greater "participation" raises considerably the income of OPEC countries, since it means that the original concessionaire's equity, now belonging to the producing-country government, has to be bought back at a price significantly higher than the "equity" price. For example, the "participation" of producer-country governments has been increased to 60 per cent in Kuwait and Abu Dhabi, to 55 per cent in Nigeria and to 51 per cent in Libya. Thus, taking the marker referred to earlier, the revenue accruing to governments from Arabian light crude, following the changes announced in November 1974, was raised to $10 a barrel.

In January 1974, Morocco increased by 200 per cent the price of phosphate rock, an important earner for that country. But this event passed almost unnoticed by the world at large. The increase in the price of crude oil, however, was an entirely different matter. The reasons for this are not hard to find:

(a) Crude oil is the most important single commodity in world commerce. In recent years it has accounted for over 50 per cent by weight of all sea-borne international trade.

(b) As far as Western Europe is concerned, oil is the chief source of energy, amounting in 1970 to 56 per cent of total energy consumption.

(c) What is more, in many applications, the most notable of which is transport, there is at present no substitute for oil products.

(d) No more than 3.5 per cent of oil consumed in Western Europe is produced domestically. Over half of Western Europe's requirements come from the Middle East.

(e) Oil accounts for a large part of the import bills of the countries of Western Europe (for example, it accounted for 11 per cent of all imports of goods into the United Kingdom in the early 1970s).

(f) Finally, the "oil economy", or the "world of oil", has always been something special. It was no great exaggeration when a spokesman for the Commission of the European Community said that it consisted of 90 per cent politics and 10 per cent oil.[3]

Hence the "crisis" has wide-ranging implications of a direct nature. But the indirect implications are even more disturbing. In the history of mankind, shortages have often developed and were usually solved by new technology; and practically all were overcome by applying massive additional quantities of energy in some conventional or new form. This was one of the chief reasons why demand for energy has escalated so rapidly. According to the exponents of the Club of Rome, who publicised their views well before the present acute crisis became apparent, this possibility was seriously jeopardised by the fundamental problem of the world nearing the final exhaustion of fossil fuel resources.[4] The 1973–74 events in the oil world have directed attention to this underlying hypothesis. But they have done so with a considerable difference. Whereas the earlier students of the question expected serious shortages of energy to develop after the turn of the century, the effect of the recent events may be that the world is facing an immediate problem, albeit on a smaller scale.

By the spring of 1974 the supply shortage had eased. It did not last long enough to cause a major world recession or have any dangerously depressing effect on West European economies, although should such a shortage become a permanent feature, either for political or other reasons, it could certainly cause more than just inconvenience. The price effects, however, will hit hard and have serious implications for the international system of trade and payments.

REAL CAUSES OF THE CRISIS

Leaving aside the role of oil as a political weapon, the question naturally arises: What were the real causes of the oil crisis? While the War of *Yom Kippur* provided the occasion for a major change in the conditions under which oil would be supplied in future, it was not – as already hinted – the cause. The festering of the Arab–Israeli problem, and the responses to it of Western governments, had influenced the attitudes of the oil-producing countries towards the Western powers. But there were two more important factors which influenced the situation:

(a) First was the determination of the oil-producing countries to obtain control over the extraction and disposition of their major natural resource rather than delegate that power to others. In 1960, at a time of surplus production, the oil companies succeeded in reducing posted prices. Irritated by this action, and alarmed at the drop in their revenues, the producing countries formed OPEC, which during the 1960s achieved only very modest gains.[5]

(b) Then towards the end of the decade there was the shift in the balance between supply and demand, largely as a result of the United States becoming an increasingly large importer, but with the Japanese market also growing out of all recognition.

To elaborate, the events of the autumn of 1973 were the culmination, in fact, of a process which began with the creation of OPEC (including the immediately preceding circumstances). In the mid-1960s the oil-producing countries unilaterally changed the terms of their agreement with the oil companies. During the 1960s, however, the *posted* price of crude oil remained stable and in real terms it actually fell. The *market* price declined, and in real terms it also fell significantly.

When the oil-producing countries began to exploit their improved bargaining position in 1970, their position was strengthened by the economic growth being enjoyed by the major industrial countries and also by the temporary interruption of piped supplies of Arabian crude oil to the ports of the Eastern Mediterranean. Under the Teheran Agreement of 1971 the "take" of the producer governments was to rise steadily until 1975. This was followed by the Geneva Agreement of 1972 which was intended to compensate the oil-producing countries for exchange-rate adjustments – but it has since been shelved.

Meanwhile, American requirements had risen to such an extent, and could only be satisfied by imports (since domestic output was only increasing slowly), that it led to the addition of unprecedented quantities of the producers' exports. In 1972–73 some of them, notably Libya and Kuwait, began to restrict exports (on grounds discussed in Chapter 6 below). Thus, the earlier concern in the United States over possible shortages of energy started to become well founded.

WIDE-RANGING IMPLICATIONS

There has accordingly begun a significant transfer of economic power from the oil-importing countries to the oil-producing countries. This shift in power will continue until such time as the more fortunate oil-importing countries can bring into production additional sources of energy to meet domestic requirements. Many of these countries are not likely to have much room for manoeuvre. For all of them, though, there will be a period when the oil-producing countries will be able to exercise considerable control over both the supply and price of oil, since the elasticity of demand is low in the short term. It is likely to be much higher in the long run because of the wide range of substitutes that can be expected to become available.

In the short run, the greatly increased prices for oil can be "lived with" by the industrialised countries, although the consequences are far-reaching. When considered as a proportion of the total income of oil-importing countries, the cost in terms of real income transferred to the oil-producing countries is of an order of magnitude that can just about be managed – albeit not without difficulty. The order of magnitude still totals many billions of dollars. Thus, the institutions and management of the international system of trade and payments cannot fail to be affected, however much the oil-producing countries see their own economic interests dependent on the orderly development of the world economy as a whole. That is why this first volume to be produced as part of the Trade Policy Research Centre's programme of studies on the Reform of the International Trading System has been devoted to *The Economics of the Oil Crisis*.[6]

In this connection might be cited a report on the economic future of the European Community, prepared under the auspices of the Institut für Weltwirtschaft in Kiel, West Germany, that was published in 1974.

> All aspects of economic life will be affected by developments in the international oil market: the direction of social and economic policies, the rate of inflation, conditions in capital markets, exchange-rate relationships, the movement of the Community towards economic and monetary union, the reform of international monetary arrangements, the maintenance of the multilateral trading system and the problems of developing countries. Failure to understand the situation fully and, on a basis of international cooperation to adopt concerted policies, could pose a serious threat to the level of world economic activity. The greatest danger facing the world economy is that individual governments will try in isolation to cope with the situation from a nationalistic standpoint.[7]

The change in circumstances will affect economic life in the ways suggested not only because of the sudden nature of the change, but also because of the magnitudes involved; a difference in degree, if large enough, can convert itself into a difference in kind. Before embarking on the following chapters which analyse the economic implications of the rise in oil prices, it would be as well to assess, if only broadly, the magnitude of the surplus revenues likely to be generated in the coffers of the oil-producing countries in the remainder of the decade.

SIZE OF THE SURPLUS REVENUES

The estimates which follow are necessarily rough since the revenues of the oil-producing countries associated within OPEC depend on the quantity and the price of oil. Production can be curtailed – for example, in order to raise the price – or it can be maintained or expanded. Similarly, the price of oil can be artificially raised further, but it can also fall. Bearing in mind these caveats, a simple estimate of the surplus revenue likely to accrue to OPEC will show the dimension of the problem facing the Western world (that is, the world outside the Soviet–Chinese spheres of influence).

Such a calculation has been used in the following tabulation which indicates how OPEC's annual surplus might change, maintaining the 1973 level of exports and a representative 1974 price. These two assumptions are admittedly unrealistic. The figures demonstrate, though, the order of magnitude involved:

	1973	*1974*
OPEC exports (million barrels per day)	28.4	28.4
Revenue per barrel ($)	2.6	9.0
Total revenue ($1,000m)	27.0	93.5
Expenditure (imports, etc.) ($1,000m)	20.0	23.5
Surplus ($1,000m)	7.00	70.00

In fact, the quantity of oil exported during 1974 remained around the 1973 level, but even higher prices and participation raised the revenue to $125–130,000m and imports amounted to $35–36,000m. The future surplus naturally depends on the size of oil shipments and their prices as well as the level of imports, possibly rising further, albeit more slowly.

Roughly one-half of the excess income of OPEC is likely to come from Western Europe. The additional burden on West European countries also depends on the quantity of oil they will purchase as well as on the price. In view of the large gains involved, however, precise figures are hardly necessary to illustrate the situation. The following table compares the 1974 excess oil cost due to the higher prices with the end-1973 reserve position for certain West European countries.

	1974 excess oil cost ($1,000m)	*December 1973 reserves ($1,000m)*
Germany (FR)	6–7	31.7
France	5–6	8.1
Italy	4–5	6.1
United Kingdom	5–6	6.3
Other Western Europe	10–12	30.0

In the case of Italy and the United Kingdom – and possibly also some smaller West European countries – the excess oil bill is nearly equivalent to the country's total reserve. In many other countries – including France – it will account for a very considerable part of reserves.

On similarly crude calculations, the excess cost of oil imports is likely to amount to around $18–20,000m for the United States, $10–12,000m for Japan, and $7–10,000m for the developing countries (with end-1973 reserves running at $13,800m, $11,600m and about $27,000m respectively). The last group excludes oil producers like Australia, New Zealand and South Africa.

Higher oil prices present a much bigger problem for the developing countries, although the position varies from country to country. Some countries have benefited appreciably from the 1972–74 commodity boom and increased their reserves to unusually high levels, but for others the new burden could be very severe indeed. For example, India's excess oil payments probably reached $1,000m in 1974, compared with reserves of some $1,400m and an export income of about $2,500 in 1972. Pakistan, Sri Lanka and quite a number of other developing countries are similarly placed. For them no other obvious solution can at present be seen than some form of aid. Their need has been recognised by the oil producers who may, according to unconfirmed reports, allocate considerable sums to alleviate the plight of the developing countries, either through established channels (such as the World Bank), through a new development bank funded by OPEC, or through some other intermediary.

Thus on the basis of early 1974 prices, and assuming the oil-importing countries are prepared to carry the debt involved, there could conceivably accumulate by 1980 balances estimated at $250–350,000m. In view of the further increase in oil prices in November 1974, and the *general* maintenance of demand, the accumulation could be significantly greater by the end of the decade. Although these balances may begin to run down in the closing years of the present decade or in the 1980s, the impact of the accumulation and the consequences of the enormous increase in the revenues accruing to the OPEC countries and of their disbursement on the pattern of world production, trade and financial flows as well as changes in them will be profound.

OUTLINE OF THE ESSAYS

The essays in the present volume fall into three broad groups: (i) those applying the tools of economic analysis to the basic implications of the rise in oil prices; (ii) those endeavouring to apply the analysis to diagnose and assess in quantitative terms the consequences of the rise in oil prices for the countries of Western Europe in particular, as well as for other countries, both developed and developing; and (iii) those concerned with the impact on international economic organisation of the changes in the international oil market.

In Chapter 2, W. M. Corden provides an analytical framework for the short-term implications of the higher level of oil prices, covering the consequences for both internal and external balance – that is, for domestic employment and the balance of payments. Dr Corden and Peter Oppenheimer discuss in Chapter 3 the purposes for which a country in deficit, whether because of its oil account or for any other reason, can legitimately borrow abroad. The view presented, that external borrowing is justified only if used to augment domestic investment, is familiar enough; but it deserves particular attention at a time when governments have been tempted to borrow for the purpose of maintaining domestic consumption.

Jan Tumlir is basically concerned with the same issues, but in Chapter 4 he takes a longer view, employing a different approach. It starts from a realistic premise that high-income countries have benefited in the past from relatively low prices for imported oil. Now with higher oil prices, the rich developed countries must increase investment in alternative energy sources, to reduce oil usage. The analysis, developed against the background of the "transfer problem of the 1920s" (in which Keynes was the leading participant), is also concerned with the international implications of the accumulation of vast sums in the hands of the oil-producing countries. Dr Tumlir presses for the surplus oil revenues to be deployed in a way that will assist a shift in low-technology industries from developed to developing countries. But to bring about such a shift, which would be in the interests of both developed and developing countries, it would be necessary for the former to open their markets to the latter.

In the second group of essays, Alasdair MacBean (Chapter 5) reviews the impact of the rise in oil prices on the trade of developing countries. Chapter 6 is devoted to a review of the energy situation in Western Europe, discussing past trends in production and consumption, estimating the likely development of supply and demand, and evaluating energy policies past and present. Chapter 7 considers the capital requirements for the development of alternative sources of energy in the major industrial countries. The conclusion is that the effort will absorb a substantial part of the growth potential of the countries concerned and, by increasing the share of investment and decreasing that of consumption, strengthen inflationary pressures.

In the final group of papers, Chapter 8 is a survey by Harry G. Johnson of higher energy costs for the world's financial structure. Then the last chapter deals with the implications for, or rather the structural changes likely to ensue in, the international trading system.

NOTES AND REFERENCES

1. It was in the United States, in the late 1960s, that the nature of the impending "energy crisis" was generally first appreciated. See the remarks of J. G. Moody, Senior Vice-President of Mobil Oil, at the 1970 Convention of the American Association of Petroleum Geologists, as reported in *Petroleum Press Service*, London, September 1970.
2. An account of the developments which led to the energy crisis can be found in Frank McFadzean, "Economic Implications of the Energy Crisis", in Hugh Corbet and Robert Jackson (eds), *In Search of a New World Economic Order* (London: Croom Helm, for the Trade Policy Research Centre, 1974; New York: John Wiley, 1974). Also see James E. Akins, "The Oil Crisis: This Time the Wolf is Here", *Foreign Affairs*, New York, April 1973,

and M. A. Adelman, *The World Petroleum Market* (Baltimore: Johns Hopkins University Press, for Resources for the Future, 1972). Attention might also be drawn to Christopher Tugendhat and Adrian Hamilton, *Oil: the Biggest Business* (London: Eyre Methuen, 1974).

3. *Energy Policy*, Guildford, December 1973, p. 263.
4. D. H. Meadows *et al.*, *The Limits of Growth*, Report for the Club of Rome (New York: Potomac Associates, 1972).
5. Abdul Amir Q. Kubbah, *OPEC Past and Present* (Vienna: Petro-Economic Research Centre, 1974) and Ashraf Lutfi, *OPEC Oil* (Beirut: Middle East Research Centre, 1968). Also see George W. Stocking, *Middle East Oil: a Study in Political and Economic Controversy* (London: Allen Lane, the Penguin Press, 1971).
6. In this connection, see McFadzean *et al.*, *Reform of the International Commercial System*, Bellagio Memorandum (London: Trade Policy Research Centre, 1974), reproduced above in the preliminary pages.
7. Sir Alec Cairncross *et al.*, *Economic Policy for the European Community: the Way Forward* (London: Macmillan, for the Institut für Weltwirtschaft an der Universität Kiel, 1974; New York: Holmes & Meier, 1975) p. 19. The report has also been published in German (Munich: Piper, 1974), in French (Paris: Presses Universitaires de France, 1975) and in Italian (Milan: Rizzoli, 1975)

CHAPTER 2

Framework for Analysing the Implications of the Rise in Oil Prices

W. M. CORDEN

In this chapter are set out some short-run implications of the rise in oil prices for oil-importing countries. The aim is to provide a framework for analysis and to draw attention to various possibilities. Its purpose is not to make predictions.[1]

To begin with it can be supposed that oil-producing countries arbitrarily raise the price of oil and then supply any quantity demanded at that price; or alternatively it can be supposed that they reduce the quantity of oil supplied and the price rises to a new equilibrium level. Provided an equilibrium situation is assumed – with no excess demand for oil at the ruling price – the two processes amount to the same thing.

Considering any one oil-importing country facing a rise in the price of oil, there will be substitution effects against oil:

(a) Other forms of energy sources, such as coal, will be substituted for oil, the extra coal resulting from additional domestic production or from imports.

(b) Other factors of production, such as capital, will be substituted for energy. Extra shipping capacity is required, for instance, if ships travel more slowly to save oil.

(c) Relatively energy-intensive products will rise in price and their consumption will be reduced in relation to less energy-intensive products.

For the purpose of the present discussion, domestically-produced oil can be regarded as an "other" form of energy, like coal, so that "oil" refers to oil imported from members of the Organisation of Petroleum Exporting Countries (OPEC).

In addition there will be an income effect. It will be assumed initially that money income and expenditure in the oil-consuming country are held constant. (This assumption will be re-examined shortly.) In that case, the rise in the price of oil will cause real income to fall. Imports of oil will fall not only because of the various substitution effects, but also because of the income effect. The net result is represented in Figure 2.1, the price having risen from *OP* to *OP′* and the quantity having fallen from *OQ* to *OQ′*.

This is the simplest first-stage analysis. The problem is to go on from here.

FIGURE 2.1

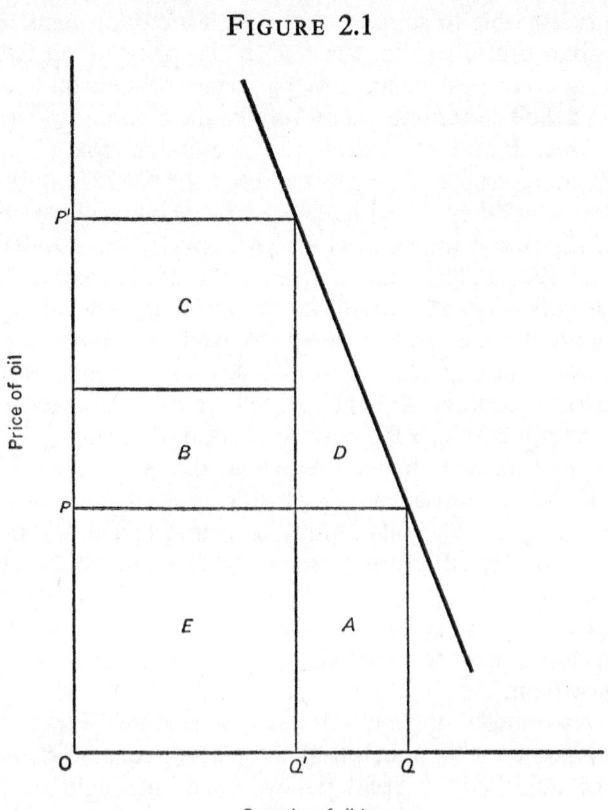

MONETARY POLICY AND INTERNAL BALANCE

The rise in the price of oil will have various domestic effects in the oil-importing countries. It might be deflationary in its impact, reducing demand for goods and services, thereby creating unemployment. There might be offsetting policies designed to maintain demand. Even when there are offsetting policies, the nature of the initial impact is of interest, because offsetting policies may operate with lags and also because one needs to know the initial impact to know what offsetting policies are required. The analysis will first treat all oil-importing countries as if they were one country. The balance of payments refers then to the balance of payments relative to the OPEC countries. (This assumption will be removed in the next section.)

It is reasonable to suppose that the elasticity of demand for oil is less than unity, so that the rise in the price of oil causes the importing countries' balance of payments on current account to go into deficit. (Assume there was balance initially.) In Figure 2.1 the areas *B* and *A* are drawn to be equal, so that *C* represents the deficit. A crucial issue now is how the OPEC countries are paid and what they do with their extra incomes. They might be paid in the importing country's own currency (e.g. dollars) or in gold, and they might choose to hold the extra income in liquid form, to invest it long term, or to spend it. The analysis will begin with the case where they are paid in dollars (or marks, francs, yen, sterling . . .) and choose not to spend them. The oil-importing country will then benefit from a short-term capital inflow exactly equal to the current account deficit.

Let us first see what happens when the dollar balances deposited by the OPEC countries simply lie idle. Initially income is equal to expenditure in the oil-consuming country. The oil price rise causes an increased proportion of this income to be spent on imported oil, and hence less to be spent on domestically-produced goods. The effect appears thus to be deflationary. If money wages are constant and profit margins cannot be squeezed there will be unemployment.

The restoration of internal balance requires expenditure to exceed income by the amount of the current-account deficit. This might be achieved by fiscal policy. Taxation might be reduced yielding a budget deficit equal to the full employment current-account deficit. The budget deficit would be financed on the capital market, the government in effect borrowing the OPEC-owned balances. There would be no need for domestic credit creation.

It is also possible, of course, that the OPEC-owned balances did not stay idle in the first place. The extra deposits might have brought down the structure of interest rates or made credit generally easier to obtain, and therefore led to additional investment or, maybe, to consumer-credit spending making up for the reduced spending out of income. Firms might borrow in order to cover losses brought about by the higher oil prices. Demand for funds may be stimulated by the increased profitability of investment in domestic oil development or oil substitutes. Furthermore, savings of non-OPEC citizens might fall, owing to the fall in their real incomes. For these reasons, expenditure would have exceeded income automatically, even without a budget deficit. One might imagine that this automatic effect would go only part of the way, some budget deficit still being needed.

If the OPEC countries choose to be paid in gold rather than dollars, the current-account deficit will not be offset by a short-term capital inflow. The initial effect will be deflationary as before; and there will be no chance of an automatic offsetting effect owing to any dollar balances deposited by OPEC countries being borrowed. Internal balance requires, then, domestic credit creation equal to the balance-of-payments deficit. The increased credit might finance a budget deficit which, as before, is brought about by tax reductions, or it might finance private investment.

The general conclusion at this stage is that the initial impact of the higher oil price would probably be deflationary, but – given present assumptions – appropriate fiscal and monetary policies can maintain internal balance.

BALANCE-OF-PAYMENTS EFFECTS

It shall continue to be assumed until the section dealing with the expenditure of oil revenue and the transfer effect, that the OPEC countries choose to hold their gains in income in gold or liquid balances. If the oil-importing countries are thought of as one country, there is no need to be concerned with the balance-of-payments effects. The "balance-of-payments problem" only comes in if a distinction is made between different oil-importing countries. Consideration will now be given to the effects on these countries separately, always assuming that each country maintains internal balance by whatever fiscal and monetary policies are appropriate.

Primary Effects

For a typical oil-importing country, say France, the oil price rise creates a *primary* current-account deficit equal to the increased value of its oil imports (the area *C* in Figure 2.1). These deficits for all oil-importing countries add up to the increase in the reserves of the OPEC countries. If the reserve increment resulting from France's deficit is held wholly in francs there will be an offsetting short-term capital inflow for France and no balance-of-payments problem (that is, no net fall in the demand for francs). But if the OPEC countries choose to hold dollars, the demand for francs will fall and for dollars will rise.

The issue is how the ratios of the various primary deficits compare with the OPEC countries' portfolio preferences. The relative demand for currencies will change. Defining the balance of payments as including both current and capital account, including short term, some oil-importing countries will go into deficit and others into surplus. The surplus countries are those where the short-term capital inflow outweighs the *primary* current-account deficit. To go back to the case of France, one could then envisage France borrowing dollars (perhaps through some recycling arrangement), the franc being devalued relative to the dollar, or France seeking to reduce or eliminate her current-account deficit through deflation. From the point of view of the oil-importing group as a whole the effects of the rise in the price of oil would be minimised if all reserve changes were automatically recycled.

In so far as the OPEC countries choose to hold gold, rather than liquid balances, and this gold comes out of central bank reserves, the type of problem just posed does not arise. But the potential problem seems more serious. All countries will go into balance-of-payments deficit equal to their primary deficits. Recycling cannot restore reserve levels. If countries place some value on their gold reserves they may seek to rebuild reserve levels at the expense of other countries, possibly by competitive devaluations or deflation. Creation of Special Drawing Rights (SDRs) on the International Monetary Fund (IMF) could solve this problem, provided these are thought adequate substitutes for gold. A rise in the price of gold will help only as long as all the central banks' gold stocks have not passed to the OPEC countries.

Secondary Effects

In addition there will be *secondary* balance-of-payments effects. These can be complex, and this discussion is not exhaustive. Broadly, there can be substitution effects, complementarity effects and income effects.

Countries which, directly or indirectly, are exporters of oil substitutes, may obtain secondary surpluses. Countries which import oil substitutes will obtain secondary deficits additional to their primary deficits. One must take all the substitution effects listed at the beginning of this paper into account. A country exporting bicycles might be regarded as exporting an oil substitute. On the other hand, exporters of goods that are complementary with oil or with energy-intensive goods will obtain deficits, while imports of such goods will fall.

It may be difficult to distinguish goods on this basis and one might argue that, apart from the actual energy-generating materials (notably coal, natural gas and a few special cases), most goods are neither clearly substitutes for nor complements with oil. One can readily think of ambiguous cases. Natural rubber, for example, is a substitute for petroleum-intensive synthetic rubber, but both are complements with oil as automobile inputs.

In addition, real incomes and expenditures are likely to fall. The nature of this fall will be discussed in the next section. For any one country this income effect will lead to both reduced imports and reduced exports. Countries which tend to export goods that have relatively low income elasticities (such as some foods) will obtain secondary balance-of-payments surpluses because of this effect, while exporters of income-elastic goods and services (such as tourism) will obtain deficits.

These secondary balance-of-payments effects are likely to have further effects: compensatory short-term capital flows, exchange-rate adjustments, upward cost–price movements by surplus countries, or deflation by deficit countries.

Until the fifth section, beginning on page 23, the various problems just discussed shall be ignored. If the OPEC countries choose to hold their income gains in gold it can be supposed that additional SDR creation makes up the reserves of oil-importing countries. If the OPEC countries choose to hold currencies, such as dollars and Deutschmarks, it can be further supposed that recycling or borrowing on international short-term capital markets solves various countries' problems. Finally, one might suppose that substitution, complementarity and income effects are so widespread and subtle that for most countries they yield no net secondary balance-of-payments effects.

INFLATION AND MACRO-UNEMPLOYMENT

Going back now to the model of the first section, and thinking of the oil-importing countries as a single country, it will continue to be assumed that the OPEC countries do not spend their income gains, but maintain them as liquid reserves.

The reduction in the supply of oil will have lowered real income and expenditure ("absorption" of goods and services) in the oil-importing country. This is true even when full employment is maintained (as we have assumed so far) and all adjustments through substitution effects have taken place. The economy has less oil available to it and has to adapt to inferior or more expensive substitutes. This *equilibrium real expenditure fall* (or absorption fall) is approximately represented in Figure 2.1 by the area $(B+D)$. This is the initial fall in real income owing to the rise in the price of oil $(C+B+D)$ *less* that part, namely C, which comes back to the oil-importing country and so is available to be spent by it. It is identical to the area $(A+D)$ which is the social valuation of oil imports $Q'Q$ forgone.

It was noted in the first section that money expenditure needs to rise in order to maintain full employement. At the same time it is now found that real expenditure has to fall. These two results can be reconciled as follows. Money income M is constant and is initially equal to money expenditure. Let initial expenditure on domestically-produced goods be R. Then $M=R+(E+A)$, where the expression in brackets is initial expenditure on oil imports. When the price of oil rises expenditure on oil rises by C to $(E+B+C)$, and on domestic goods falls by C. To restore demand for domestic goods, total expenditure is then increased by C which is assumed to be spent wholly on domestic goods. We end up with the same total resources engaged in production of domestic goods and services as before (though, no doubt, a different output composition) and lower oil imports. So real expenditure is lower than it was initially. The rise in the price of oil has more than offset the increase in money expenditure.

With total real expenditure having to fall, it is likely that the after-tax real wage compatible with full employment needs to fall. This assumes that any squeeze on profits and investment would not be sufficient to absorb the required real expenditure fall. This required fall in the real wage *could* take the form described above: a rise in prices owing to the higher oil price combined with constant money wages. Alternatively, money wages might rise, but slower than prices. But it is more likely that the price rise brings forth money wage increases designed to maintain real wages, so causing prices to rise further. This price–wage spiral might be superimposed upon an inflation that would have taken place even in the absence of the rise in the price of oil.[2]

Finally, in an attempt to slow up this spiral governments may pursue deflationary policies which create unemployment. To distinguish this from the *micro*-unemployment which we shall introduce later, it may be called *macro*-unemployment. It leads to a fall in real output, and hence real income and expenditure, which is additional to the *equilibrium* fall that started the process off. If one measures a "recession" not by unemployment but by the fall in output, there are then two reasons why reduced oil imports may cause a recession. In the next section we shall introduce a third reason.

LABOUR IMMOBILITY AND MICRO-UNEMPLOYMENT

A rise in the price of oil raises the prices of oil-intensive products and so shifts demand away from them. This reduces demand for the labour employed in oil-intensive activities. In the short run much of this labour may be immobile, so that unemployment of a "structural" nature results. The ramifications of this effect can be very great and may dominate the short-run effects of a large rise in the price of oil. In industry labour is directly complementary with energy. More generally, labour is complementary in final consumption with the motor car (garage attendants, mechanics and so on) and with oil-intensive products. If the relevant labour is truly immobile this problem cannot be solved by internal balance policies of a *macro* nature.

The immobility of such labour complementary with oil brings about a further loss of output since, if the labour had moved out of the oil-intensive activities or the activities complementary with oil-intensive products, it would have added to output elsewhere. This loss of output resulting from *micro*-unemployment will bring about a further fall in equilibrium real expenditure. If there is an attempt to maintain real wages it will strengthen the inflationary effect of the rise in the price of oil and, for this reason, may provoke policies that generate more *macro*-unemployment.

In a fundamental sense the cause of micro-unemployment is the same as the cause of macro-unemployment. Both result from wage rigidity. The chain of reasoning is that a rise in the price of oil raises the price of the oil-intensive product, reduces demand for it, and so reduces employment in that activity. But if wages were flexible downwards, unemployment would force the wage down and restore employment through two mechanisms: it would be profitable to substitute labour for oil to produce a given amount of product, in so far as such substitution is possible; and the price of the final product would rise less (the lower cost of labour offsetting somewhat the higher cost of oil), and hence the fall in demand for the final product would be modified and, in the extreme case, completely avoided.

Thus full employment might be maintained even though labour is completely immobile. In time the lower wage would encourage some movement out of the industry. The simplest type of neoclassical model, which underlies the analysis in preceding sections, assumes flexibility of relative factor prices and so does not allow for such micro-unemployment.

EXPENDITURE OF OIL REVENUE AND THE TRANSFER EFFECT

Let it be assumed now that the extra oil revenues of the OPEC countries are spent by them on goods and services produced by the oil-importing countries – rather than being hoarded or invested in short-term assets. The higher import bill of the oil-importing countries will then be balanced by an equivalent increase in exports. The balance of payments on current account relative to the OPEC countries will stay in equilibrium. Nevertheless from a short-term point of view, such a development is hardly to be welcomed by the oil-importing countries. To make way for the extra exports their equilibrium real expenditure will fall even further than before.

Transfer Effect

For the moment the oil-importing countries will still be treated as if they were a single country. Reference is made again to Figure 2.1. Assuming full employment, the loss inflicted on this country has now two parts: first, the real resources loss $(A+D)$ resulting from the reduced supply of oil and, secondly, the loss C, which is the value of extra exports to the OPEC countries, and is equal to the decline in domestic absorption of domestically-produced goods that is required if internal balance is to be maintained. There has been a transfer of real expenditure from oil-importing countries to OPEC countries so that C represents the *transfer effect.*

Money income will now be equal to money expenditure, and (if money wages stay constant) both can stay constant when the price of oil rises. The deflationary domestic effect of the additional expenditure on oil imports – equal to the extra import value C – will be exactly equal to the inflationary effect of the extra export demand. The higher price of oil brings about automatically the required reduction in *real* expenditure – that is, the required disabsorption.

The costs of *micro*-unemployment could be incorporated in the real resource loss $(A+D)$. It measures the total social value of the real expenditure reduction resulting from the reduced supply of oil, allowing for both direct and indirect effects. The effects of *macro*-unemployment are of course additional. The general argument has been that the greater the equilibrium real expenditure fall resulting from the oil price rise the greater the inflationary pressure is likely to be, and hence the greater the macro-unemployment resulting from attempts to counteract this pressure. Since the transfer effect reduces equilibrium real expenditure even further, it will increase inflation and unemployment.

Balance-of-payments Consequences

The point has been reached in the analysis to distinguish again between different oil-importing countries in order to see what happens to their balances of payments as a result of the transfer effect. For the group as a whole the balance-of-payments effect will be zero. Internal balance policies are assumed for each country. Money wages and exchange rates are constant. First, then, the analysis concentrates on the transfer effect, assuming the various effects discussed in the second section (substitution, complementarity and real income effects) have no net effect on balances of payments.

Each country makes a transfer of spending power from its own consumers and investors to the OPEC countries. This will alter the relative demands for the products of different oil-importing countries if the spending patterns (the marginal propensities to spend on different goods) differ between the latter countries and the OPEC countries. If France would have spent 80 per cent of the transferred funds on French goods and 20 per cent on German goods, while the OPEC countries spend 50 per cent on each, the result will be to give Germany a surplus and France a deficit. This effect may, of course, be offset by a similar bias of Germans for their own goods. The traditional analysis of the "transfer problem" is relevant here. There is a general presumption that each oil-importing country has a higher propensity to spend on its own goods relative to foreign products than the OPEC countries have. It follows that the larger is a country's transfer relative to other countries, the more likely is it that it will end up with a payments deficit. In addition one must, obviously, take into account the special demand patterns of the OPEC countries.

Changed Pattern of World Demand

This analysis can next be combined with the conclusions of the second section. Each country has a *primary* payments deficit equal to the increased value of its oil imports. If the OPEC countries spend the extra income, this is then transferred in the way described here. The net result of its transfer and the transfer of all the other oil-importing countries is to transform this primary deficit into a *post-transfer* deficit or surplus. The transfer of its own deficit is likely to modify, although not eliminate, the primary deficit. But when its increased exports owing to other countries' transfers are taken into account, it may end up with a post-transfer surplus. In addition there are *secondary* balance-of-payments effects as described in the second section.

A country which at this stage ends up with a balance-of-payments deficit will have a level of money expenditure that exceeds money income by the amount of the deficit. In real terms, countries with deficits will gain relative to countries with surpluses; inflationary pressure will be less for the former than the latter. This of course is only a short-term gain for the deficit countries, for they must be running down their foreign assets or adding to their liabilities. But the relationships are likely to be reversed once allowance is made for exchange-rate adjustments designed to restore payments equilibria. If the net result of the transfer effect and of the secondary effects has been to shift the world demand pattern towards Germany, then German prices will be able to rise relatively, and German real expenditure will be able to increase relative to what would have happened otherwise. Germany will still suffer from the various losses described earlier, but these will at least be modified by this new consideration. On the other hand, countries which suffer from a demand shift away from their goods will incur yet another form of loss, in fact a deterioration in the terms of trade relative to other oil-importing countries.

CONCLUSIONS

This discussion has not been comprehensive. One might allow for: (i) the effects of the OPEC countries investing their gains on the stock exchanges of the world, which would bring about various portfolio adjustments, but perhaps in the short term would not be so different in its effects as the investment in liquid balances; (ii) the effects of the higher oil price and reduced quantity on oil company profits, taking into account the use and distribution of these profits; (iii) a situation where the oil price is below equilibrium, given the quantity restriction by OPEC countries, even though it may still have risen quite severely (leading to rationing by companies or governments, or both, which may inflict distinct resource losses compared with rationing by price, even though there may be a gain owing to a reduced transfer effect); (iv) income distribution effects within oil-importing countries, notably the income shifts towards producers of substitutes (including those oil companies depending mainly on non-OPEC oil); (v) reduced taxes on oil imports in non-OPEC countries, so modifying or avoiding various substitution effects and domestic income redistribution effects; (vi) OPEC countries shifting their reserves in and out of currencies according to whim or for political motives (which would impose serious adjustment costs on the oil-importing countries over and above all the other costs, unless they have the sense to offset these currency shifts with adequate recycling arrangements). Finally (vii) one might discuss optimal policies, whether for the OPEC countries, the oil companies, or the oil-importing countries.

To conclude, it might be useful to list the various short-term costs and benefits to oil-importing countries resulting from the rise in the price of oil:

(a) The minimum real resource loss directly resulting from the reduced oil supply. It is the cost which would result if there were not unemployment and it is represented by the area $(A+D)$ in Figure 2.1. It requires that equilibrium real expenditure fall.

(b) The output loss owing to macro-unemployment resulting from counter-inflationary policies. The cause of the inflation is the fall in equilibrium real expenditure. (One might also add here the costs of inflation *per se*.)

(c) The output loss owing to micro-unemployment. This reduces equilibrium real expenditure further (and could be included in $(A+B)$).

(d) The transfer effect, which operates only if the OPEC countries spend their gains. It also reduces equilibrium real expenditure (represented by C in Figure 2.1).

(e) The loss *or gain* owing to changed terms of trade relative to other oil-importing countries. This results from the consequences of the transfer and the various secondary effects.

Mention should also be made of two possible costs referred to in the preceding paragraph, namely

(f) The adjustment costs when the OPEC countries hold foreign exchange and switch their portfolios, and

(g) The costs of rationing.

(h) In addition, if the OPEC countries choose to hold gold and the oil-importing countries do not substitute SDRs, there may be various costs of reduced liquidity or measures to restore liquidity.

Long-term effects, such as the consequences of increasing financial power for OPEC governments, have not been discussed in this chapter. Some of these are pursued in Chapter 3.

NOTES AND REFERENCES

1. This chapter is based on a lecture given at the Graduate Institute of International Studies at the University of Geneva on 20 December 1973 which was published as W. M. Corden, "Implications of the Oil Price Rise", *Journal of World Trade Law*, London, March–April 1974.
2. The discussion here has assumed implicitly that the increase in domestic money expenditure (the area C) is available to be used for consumption in the oil-importing countries. For example, the government might borrow the OPEC funds and use them to subsidise food, or to reduce taxation on wage-earners. In the short term they are indeed likely to be used to a substantial extent for the maintenance of consumption. In so far as they are used for investment, the after-tax (and subsidy) real wage needs to fall even more than indicated here, and the price–wage spiral is likely to be intensified. The case for using at least some of these funds for investment is developed in the next chapter which focuses on longer-term issues.

CHAPTER 3

Economic Issues for the Oil-importing Countries

W. M. CORDEN and PETER OPPENHEIMER

The dramatic increase in oil prices towards the end of 1973 has posed major problems for economic policy in both oil-producing and oil-consuming countries.[1] The oil producers, members of the Organisation of Petroleum Exporting Countries (OPEC), have to decide where and how to hold the surplus revenues accruing to them. They also have to work out future policy on oil production and on the diversification of their economies. These problems were beginning to appear even at prices ruling before October 1973. They have since been greatly magnified. Consumers, for their part, while pressing ahead with the development of alternative energy sources, have to manage the payments deficits which are the counterpart of the producing countries' surpluses.

The object of this paper is to outline as concisely as possible the main issues arising from these deficits, especially – although not only – as they relate to the United Kingdom. Some of the issues have of course been widely recognised.[2] Others have been comparatively neglected. What it is intended to bring out here is the crucial distinction between borrowing to finance consumption and borrowing for investment. It is a distinction that does not appear to be sufficiently appreciated. Immediately after the oil crisis was precipitated, it was understandable that official and public attention should focus on the impact of higher petroleum prices on inflation and the level of economic activity, but several months later those aspects still seem to be the focus of attention.

In the first two sections are examined the implications of higher oil prices for aggregate demand and employment in the major oil-consuming countries. A brief look in the third section at the demands being made on financial markets provides a bridge to the fourth section, which tackles the central question of consumption *versus* investment. The final section examines whether and how Britain's present position is altered by the prospect of North Sea oil.

ADJUSTMENT OF AGGREGATE DEMAND

Let us begin by asking whether the rise in the price of oil raises or lowers the demand for goods and services produced by the non-OPEC countries. The oil price rise is, in the language of economists, a tax collected by OPEC countries from oil users, direct and indirect.[3] It leads to some fall in the quantity of oil consumed (relative to what would have happened otherwise), but since the demand elasticity response is low, expenditures on oil go up. Hence expenditures available for buying the products of the non-OPEC countries would appear to go down. It is assumed at this stage in the analysis that countries' monetary and fiscal policies, money wage rates and profit mark-ups remain as they would have been if oil prices had not gone up. On this line of argument it is widely claimed that the effect of the oil price rise is demand-*de*flationary and calls for deliberate fiscal measures to avoid a recession.

But this is not the whole story. An obvious offsetting effect is that some part of the extra incomes of the OPEC countries will be spent on imports from the non-OPEC countries. Additional export demand will thus make up some of the aggregate demand gap. These exports may build up quite rapidly over time. Nevertheless a huge gap remains.

There is, though, a further offset. The OPEC countries must deposit their surplus revenues somewhere so that – for the non-OPEC world as a whole – a capital inflow equal to its current account deficit with the OPEC countries is inevitable. The question then is how these increased capital funds affect aggregate demand. Clearly the funds will not all stay idle. To the extent that they are lent out and lead to extra private investment spending they have a stimulating effect on aggregate demand. In the limiting case one might imagine all the funds leading immediately to extra investment outlays. In that case the net result of the oil price rise will have been to keep aggregate demand constant. It is no doubt more realistic to suppose that some of the funds are lent out and spent, while others stay idle. There is therefore another important offsetting effect to the initial deflationary impact which ought not to be neglected. But on balance the effect – at least for the non-OPEC countries as a whole – remains deflationary.

What in effect has happened is that the world propensity to save has gone up, owing to a redistribution of income towards high savers (the oil producers). For the maintenance of aggregate demand the extra savings must either be offset by lower savings elsewhere or be converted into investment. The choice between these alternatives is the subject of the fourth section of this chapter. Here it is simply assumed that extra savings need to be matched by extra investment. The question then is how far the private sector will be induced by even a zero or negative real rate of interest to undertake the additional investment; and how far government fiscal action will be required. This is the typical Keynesian problem.

On the one hand, investment opportunities will have been increased by the need for oil substitutes; on the other hand, it is doubtful whether *sufficient* investment opportunities can be generated reasonably quickly to absorb the extra savings. In the course of time, investment financed directly or indirectly by oil funds will become an increasingly important source of demand, even though it may be inhibited in the early aftermath of the oil crisis by reduced profit expectations and the uncertain economic outlook. Hence the extent of fiscal offsetting measures should be reduced over time.

Another way of putting the matter is to say that the non-OPEC world's money supply has remained unchanged, but if some of the funds stay idle, its velocity of circulation will fall – at least in the short run – unless governments step in to borrow and spend some of the funds, so restoring the velocity.

So far the situation has been analysed from the point of view of the non-OPEC countries as a whole. There is then no balance-of-payments problem, apart from the need to ensure that producers' willingness to continue selling oil is not undermined by some crisis of confidence in financial markets. This aspect is touched on in the third section. Meanwhile, the matter looks different if any one country, such as the United Kingdom, is considered on its own.

Initially each single country will acquire a current-account deficit (or a cut in its current-account surplus) attributable to the oil price rise. This worsening of the current account does not just equal the extra value of oil imports. Additional exports to OPEC countries must be subtracted. Various indirect effects of the oil price rise should also be taken into account, at least in principle. For example imports of oil substitutes may go up. All in all however most countries will wind up with a deficit on current account. At the same time, a country may be a recipient of surplus oil funds; in Britain's case, the inflow of such funds is large. Bearing in mind that some of the funds may quickly flow out again, there will be a net effect on the capital account. Some non-OPEC countries may thus find themselves in surplus and some in deficit on current and capital account combined. In principle the sum of all surpluses (including OPEC) is equal to the sum of all deficits, though it may be difficult to establish this equation in practice, because of the non-uniform treatment of liquid assets and liabilities in most balance-of-payments accounts.

Consider a case, such as Britain's, where an overall (oil-related) deficit remains. If the whole of the funds that did flow into Britain financed extra investment spending, then the overall balance-of-payments deficit would be equal to the aggregate demand deficiency. If some of the funds remained idle the deficit would be less than the demand deficiency. It is certainly conceivable that a government might restore aggregate demand without altering the exchange rate. In the first instance, it would finance a budget deficit by borrowing the idle funds which have already flowed into Britain, and then it would allow itself an additional budget deficit financed by borrowing from abroad some of the oil funds originally deposited in New York or on the Eurodollar market. Internal and external balance would then be restored, if external balance is defined as equality between the current account deficit and the net capital inflow, whether borrowed privately or publicly.

The central issue is whether such borrowing is desirable and, indeed, at a limit, whether it is possible. For the moment however let it be supposed that the British government does not now wish to borrow to that extent. Hence a deficit and a demand deficiency remain. It is at this stage that the exchange rate comes in. A depreciation will in principle shift demand away from foreign goods and on to domestic goods, so simultaneously improving the current account and restoring demand for domestic goods to the desired level. Such a strategy will be feasible to the extent that other non-OPEC countries are willing to countenance the depreciation and do not offset it by engaging in a similar policy themselves. Also of course the effectiveness of depreciation depends on the availability of unused productive resources. If the lack of demand was initially remedied not by depreciation but by fiscal expansion, so inevitably worsening the current account further, a subsequent depreciation would have to be associated with a fiscal contraction.

OIL PRICE RISE THE CAUSE OF RECESSION?

The theme of the argument thus far has been that the oil price rise is demand-deflationary in the first instance, although, when account is taken of the effects on private investment spending of the ready availability of capital at low or even negative real rates of interest, the total deflationary effect may not be as large as is sometimes feared. In any event appropriate fiscal policy, associated if necessary with exchange-rate adjustment, can counteract the deflation. This leaves out of account problems of time-lags and short-term adjustment and misjudgements, with which we shall not concern ourselves here. The essential point is that recession – as the term is usually understood – is not an inevitable consequence of the oil price rise. Governments need not have it unless they want it. To that extent the situation is certainly different from the 1930s.

It is inevitable that the oil price rise affects real wages and real profits even at full employment; and if one defines a recession as a fall in real incomes or the rate of growth of real incomes rather than as a situation of excessive unemployment, then a recession is what is on the cards. But this is not the sense in which we are using the term here. Here we are only concerned with deficiency of aggregate demand.

The above argument must be qualified in one way. Any prolonged recession is likely to be government-induced and deliberately designed to restrain wage inflation. This problem had arisen before October 1973. It was clearly apparent early in 1973 that by 1974 many governments would wish to practise tight demand policies to inhibit inflationary pressures. How does the oil price rise affect this situation? It does so in two ways.

First, it produces – as we have discussed – a natural deflationary effect, so that less deliberate deflation by fiscal or monetary policy is required than would have been needed otherwise. In so far as governments underestimate the deflationary impact of higher oil prices when devising their own policies, the total deflationary pressure in the world system will of course be increased.

Secondly, the oil price rise lowers real wages for given money wages and creates, accordingly, additional pressure to raise money wages to offset this inevitable effect. While it is demand-inflationary, it has thus intensified the cost-inflationary spiral, which in turn strengthens the case for counteracting demand deflation. Using the Phillips curve concept, one could say that the oil price rise has shifted the curve in an unfavourable direction, just as a successful incomes policy would shift it in a favourable direction. One may add that if the government borrows abroad to sustain real wages in some form or other – for example borrowing to finance food subsidies or a cut in indirect taxes – it is likely for the time being to reduce the cost-inflationary pressure and hence the need for recession-biased policies. Even if the government thinks such a policy well-advised however the international financial community may take a different view and allow the government to borrow only on increasingly onerous terms or in limited amounts; and the government's freedom of manoeuvre will be narrowed. Moreover the government may be reluctant to aggravate domestic cost-inflation by devaluing, especially when the desired impact of devaluation on the volume of exports and of competitive imports is likely to operate with a considerable time-lag.

For these various reasons, the oil price rise has made government policies more recession-biased than they would have been otherwise.

DIFFICULTIES FOR FINANCIAL MARKETS?

There are in any case significant technical problems arising from the sheer volume of OPEC funds that need to be channelled through the financial system to willing borrowers. Four different problems have either become apparent already or are immediately foreseeable.

First, many of the Arab depositors showed a strong initial preference for short-term deposits; caution is leading them only reluctantly into longer-term investments, including property and ordinary shares. But there is no equivalent demand for short-term funds. Banks find themselves borrowing short and lending long on an excessive scale, and will be reluctant to take on unlimited short-term deposits. Continuing imbalance of this kind must get reflected in the term structure of interest rates. When rates obtainable on short-term deposits fall well below long-term ones, and perhaps before this happens, the Arab depositors will no doubt alter their requirements. The initial problem is partly the result of uncertainty about prospective rates of inflation and partly another lag in behavioural adjustment to a new situation.

The second difficulty is that even the largest banks are under-capitalised to handle this vast flow of business; prudence and official regulations require a minimum ratio of bank resources to deposits. Given time one might also expect this problem to resolve itself. Once the Arabs are willing to invest long term and on the stock market it will be possible to find funds readily to expand the capital of the major banks and to start new banks. Furthermore governments can step in as financial intermediaries.

But until this second problem has been brought under control there is always the possibility of a sudden crisis and of bank failures as funds are moved from bank to bank. The seriousness of this hinges on the willingness of central banks to act as lenders of last resort. Since the deposits must go somewhere, and there are not so many countries and so many banks to which they can go, it should not be beyond the wit of man – especially central banking man – to maintain adequate recycling arrangements. We believe that this could be a short-term problem, especially after the spate of losses from in-house foreign-exchange speculation by some important banks in 1973–74, in one case actually obliging the bank to close its doors. But it ought not to be a fundamental problem.

A third difficulty is that if a substantial part of the funds is going to be borrowed by a small number of governments, the risks to private banking intermediaries, even when adequately capitalised, will be immense. It would seem more sensible for large-scale borrowing to be handled directly, government-to-government, or through official international agencies. This problem is not resolved merely by inducing the OPEC countries to invest their funds at the long end of the market.

Finally there is the fundamental problem which shall be returned to in a moment. Is the demand for funds going to be sufficient, or will the banks find themselves eventually with funds that no-one wants to borrow? Presumably there is *some* negative real rate of interest at which the banks will not end up becoming involuntary holders of financial inventories; with expected inflation at continued high rates it will be possible to make this compatible with positive nominal rates. But this fundamental and essentially long-term problem could manifest itself as a short-term technical banking problem.

Subject to one qualification all but the last of these are essentially short-term and technical problems. If not fully anticipated they may prove temporary stumbling blocks, but they should still be quite promptly resolved, especially as the remedies are rather obvious and in the mutual interests of all concerned. In the absence of gross follies of national government there is no reason why difficulties of particular banks, or even of the whole capital market, need lead to recession or a repetition of the 1930s' situation.

The one qualification is that if the depositors find no secure way of depositing their funds and obtaining a reasonable return (that is, not too negative in real terms), then the flow of oil might again be restricted. If the restriction is sudden and large, the effects would certainly be severely adverse, although even here some informed opinion is that over the longer term the world can adjust without too many difficulties to reduced oil supplies from the Middle East.

BORROWING FOR CONSUMPTION OR INVESTMENT?

We now come to the fundamental issue. The main point seems rather obvious, but there are indications in much published and official comment in Britain that it is being overlooked, or at least insufficiently stressed.

The funds that the OPEC countries are putting into the world's capital markets are being *lent* to the non-OPEC world. They are not being given back as a present. If the government of a country borrows these funds in order to sustain national consumption levels – as the United Kingdom is doing not only when its public sector bodies borrow on the Eurodollar market, but also when it finances a budget deficit with sterling Treasury bills bought by OPEC governments – it will incur a growing foreign debt burden, a charge on future generations. It is only sustaining consumption levels temporarily. It is really *anticipating* consumption, something that *may* be justified by a country's rate of time-preference and expectation of future growth, but should be done consciously.

In the very short term it may be justified for countries such as Italy and Britain to borrow publicly for the maintenance of consumption in order to make a necessary adjustment of consumption more gentle and avoid transitional dislocations, notably those involved in adding to the wage–price spiral. Furthermore there may be some case for borrowing for consumption where a country expects future growth in its *per capita* incomes and wishes to smooth out the path of consumption over time. The right amount of borrowing for these purposes will be higher the lower the world real rate of interest; and this rate of interest has probably been lowered as a result of the oil price rise. But taking into account the large volume of the funds involved it seems to us that substantial borrowing for consumption cannot go on for any length of time. Indeed even if countries wanted to borrow for long periods to sustain consumption, they would in due course reach lending limits. The possibility of default would be too great.

In the main the funds will have to go wherever the investment opportunities exist. It is necessary to find investment opportunities yielding real rates of return equal to the rates of interest being charged, and thus in effect real rates which are high enough to induce continued oil production. It is well-known that the producers' decision as to whether to produce now or not depends on the expected rate of change in the real price of oil relative to the real rate of interest. Both could conceivably be negative, but if the latter is below the former it will pay to contract production.

If we want to know which countries will be able to run current-account deficits through obtaining oil funds, either directly or through the capital market, we must know primarily where the profitable investment opportunities – whether private or public – will be. Over and above this, countries may for limited periods borrow to anticipate consumption, and thus run an extra deficit, but this can hardly last. There is certainly no case for maintaining a current-account deficit equal to the "oil deficit", meaning by this the deficit that was created in the first instance by the oil price rise. A country with good investment opportunities may be able to run a bigger deficit and some countries will certainly need to run smaller ones.

This argument has an important corollary. It is at the outset mistaken to focus policy on the current account of the balance of payments or on some arbitrary current-account target. The focus needs to be on the search for investment opportunities. From the point of view of one country it is a matter of getting real rates of return at the margin at least equal to the real rates of interest that have to be paid internationally. Of course it is social and not private rates of return that are relevant, but the term "social" must not be interpreted in too liberal a way. The need is for a social return which will contribute, directly or indirectly, to servicing a debt payable in actual foreign currency.

We are of course not arguing that the balance of payments should be ignored. When a country runs a deficit which is considered to be a problem it is presumably either running down its reserves involuntarily or engaging in unplanned borrowing. All we are saying is that borrowing should be planned and rational. If it is borrowing for investment with a good rate of return it is rational; if it is borrowing for investment with a poor or negative rate of return or for consumption it *may* still be rational – but it may not. The fact of a current-account deficit does not tell us what the borrowing is for. Only in the case where borrowing is for unprofitable investment or excessively for the maintenance of consumption can one say that there is a balance-of-payments "problem" – although it is really just a problem of unwise borrowing.

It follows then that a country such as the United Kingdom should determine its optimal borrowing levels in the light of these considerations. This will then determine the current-account deficit that represents "external balance". The sum of net planned borrowing by the public and private sectors will equal this deficit (including in "net borrowing" any rundown of liquid assets). The role of the exchange rate is to switch the demand pattern between foreign and domestic goods so as to maintain full employment in conjunction with the chosen deficit (or surplus) on current account.

All of this looks at the matter from the point of view of one relatively unimportant country, such as Britain. Suppose that the sum of all the borrowing that countries want to do falls short of the funds available on the world's capital markets. This is the fundamental Keynesian problem of a shortage of investment opportunities. The funds cannot just stay idle, if full employment is to be maintained. *Someone* must either borrow them directly or engage in the Keynesian remedy of a compensating expansion of domestic credit through deficit-financed public expenditures which effectively means that the funds get borrowed by the government. But a government that does this must not forget the foreign debts it will have to pay: either the public expenditures must lead, directly or indirectly, to the buildup of assets yielding an appropriate return, or the immediate gain in consumption must be thought to warrant the corresponding loss of consumption in the future. From the point of view of Britain it needs to be stressed that a country which borrows in excess of its domestic investment opportunities in order to help maintain world levels of aggregate demand is making a sacrifice for other countries.

One might speculate for a moment as to whether it is really possible for some of the funds accumulated by the OPEC countries not to find a home anywhere because no one – private or government – wants them. The fear has sometimes been expressed that this might happen.

Suppose in the first instance that the funds are accepted by private banks in New York and that no non-American private firm or government chooses to borrow them. Non-American governments deal with their initial deficits and deflations created by the higher value of oil imports from OPEC countries by devaluing, and if they all do, then they are in fact all devaluing relative to the American dollar. Assuming the United States is indifferent to the size of its current-account deficit, it then needs to increase expenditure to maintain internal balance – since both its initial oil-induced deficit and the subsequent foreign devaluations will have reduced demand for its own resources. This increased expenditure would be directly or indirectly financed by the OPEC funds. For example it might take the form of a United States budget deficit financed by treasury bills bought by American private banks.

The improbable situation one needs to envisage is where even American private banks or the United States administration refuse to take any more funds – even when, say, zero or negative rates of interest are being paid. What then? One possibility is that OPEC countries would then impose an embargo on their exports to force consuming countries into providing adequate monetary arrangements. Alternatively the world might slide into a regime of inconvertibility.

In the latter case OPEC countries would have to ask the buyers of their oil not to make payments in a foreign currency, such as sterling or dollars, since a home for such deposits could not be found (not even through Swiss intermediaries!), but to keep them in the buyers' domestic currencies. Since Italy for example could not refuse lira deposits belonging to OPEC countries which are payments for oil bought by Italy (for otherwise the oil would stop coming), the OPEC funds will in fact have found homes. These homes will have been tied directly to the sources of OPEC income. We certainly do not expect this situation to arise. We expect that investment possibilities will emerge and that the world capital market will work, if somewhat imperfectly. Furthermore there will be for some time no shortage of governments wishing to borrow for consumption maintenance.

There is one final point on investment possibilities. In 1974 nominal rates of interest have certainly been high, but real rates – on the basis of general inflationary expectations – have been close to zero or even negative. It is of course uncertain how long such a situation will last. While it does, though, it means that the rates of return required from suitable investments of oil funds are rather modest – namely near-zero or negative after allowing for depreciation and price increases. Countries have to build up assets to match the borrowed funds – just using them for consumption is not the same – but subsequently they need to do little or no more than keep the real value of the assets intact. To the extent that market real rates are negative, even this is unnecessary. Countries can either allow assets to decline in real value or maintain them constant and borrow additionally for consumption.

WILL NORTH SEA OIL SAVE BRITAIN?

Finally, is it reasonable for Britain to borrow for consumption because of the profits which by 1980, or thereabouts, she expects to be making from North Sea oil? And should she borrow to finance investment in North Sea oil development? The answer to the second question is obvious. Provided a proper calculation is made, there is no reason why either government or oil companies should not borrow for this purpose, as indeed the latter have been doing. The answer to the first question is also simple. As the flow of North Sea oil is likely to yield substantial gains to Britain compared to the present position where Britain is importing almost all her needs at the high post-1973 prices, there is a case for *some* borrowing for consumption now. This is the argument mentioned earlier that a country may be justified in anticipating consumption to some extent if future growth is anticipated. We are here of course referring to total consumption of all goods and services, not just of oil.

The real issue is whether there is a case for borrowing to the extent of *maintaining* consumption levels that existed before October 1973. Is Britain justified in borrowing for consumption to the extent of avoiding any significant drop on 1973 consumption levels, on the grounds that such borrowing would only be needed until North Sea oil is flowing and that the interest and debt payments can come out of the eventual gains from North Sea oil? Such a case appears to us very weak, and indeed on the basis of present expectations and price and export prospects, unsustainable.

Suppose first of all that the North Sea will make the United Kingdom just self-sufficient in oil. The cost of the oil to the nation will be the price excluding tax and royalties collected by the British government. This cost will include all the current resource costs, interest payable on borrowed funds, dividends going to foreigners and also dividends going to British shareholders where these represent a normal return on capital. In so far as there is a "rent" element in the payments to British citizens the economic cost is somewhat less. Now while the tax and royalty share will be large, the remaining cost will still be greater in real terms than the price of imported oil before October 1973. The country will be better off than it is *today*, but it will not get back to where it was before 1973. One should be under no illusion about the sums that will be collected in tax and royalties. Assuming that the United Kingdom does not become a major net exporter, they will be less than the extra taxes (on top of the pre-October 1973 price) that the OPEC countries are now collecting from Britain.

It might also be added that the net cost of North Sea oil as described above *under*states the cost of replacing formerly cheap Middle East oil, because one must also take into account the excess cost of expensive oil substitutes, notably coal and nuclear energy, and the excess cost of adapting to a less fuel intensive pattern of both production and consumption. Against this there may be some offsetting gain from the fact that self-sufficiency in oil will lower Britain's average propensity to import, thereby possibly improving her terms of trade (compared with what would otherwise have happened). But as Britain is now a fairly small country in world markets, no very great improvement can be expected. On balance therefore these various additional items strengthen the argument that, for the foreseeable future, the United Kingdom will not get back to the situation prevailing before October 1973.

The above argument assumes that North Sea oil, together with alternative energy sources, will make Britain self-sufficient. If she becomes a net exporter there will be the possibility of a gain compared with 1973 (in addition to any effect on the terms of trade). The gain from exports will be measured by the tax and royalties on exports and this gain would have to be subtracted from the net excess cost of replacing Middle East oil for British consumption. There are no indications to date that the gain will be so large as actually to make the United Kingdom better off than in 1973. For it is unclear to what extent Britain will become a net exporter of oil – this depending among other things on rates of tax and royalty yet to be fixed. But of course it is possible that she will be better off than in 1973 and the greater the expected national profit (primarily in the form of tax and royalties) the stronger the case for *some* borrowing to sustain consumption in the meantime.

NOTES AND REFERENCES

1. The first version of this chapter was published as W. M. Corden and Peter Oppenheimer, *Basic Implications of the Rise in Oil Prices*, Staff Paper no. 6 (London: Trade Policy Research Centre, 1974), a revised version later appearing in *Moorgate and Wall Street*, London, Autumn 1974. Most of the revisions in the present version are in the fourth and fifth sections, dealing with British policy, where the earlier presentation of the arguments gave rise to misunderstandings.
2. Some of the issues have been examined in Chapter 2 of the present volume. Reference might be made to some of the analyses listed in the Selected Bibliography below.
3. For a discussion of the oil price increase as a tax, see Jan Tumlir's treatment in Chapter 4 below.

CHAPTER 4

Oil Payments and Oil Debt and the Problem of Adjustment

JAN TUMLIR

It is still uncertain whether the world economy can adjust to the sharp rise in oil prices without a major loss of output. It is, in other words, uncertain whether all the main aspects of the new situation are sufficiently understood. This chapter deals with four of these aspects: (i) the transfer mechanism, (ii) the nature of the payments problem, (iii) the implications for the oil-importing countries of the large external debt they will have to carry and (iv) their structural adjustment to the higher level of oil prices.[1]

The analysis has been simplified to the bone. It is focused on the "real" magnitudes and admittedly neglects many pertinent monetary and financial considerations. In the present case there is some danger that the "veil of money" could obscure the very issues that it is this chapter's purpose to elucidate.

OIL PAYMENTS IN THE THEORY OF INTERNATIONAL TRANSFERS

The first question posed by the increase in the price of oil is the following. How are the oil-importing countries to pay the oil-exporting countries? There are a number of ways of making the payment. But to see clearly where each leads, it is necessary to proceed on the basis of some more general theory, one into which the post-1973 situation can be fitted.

Fortunately such a theory exists ready-made, known to students of international trade as the "transfer problem". It derives from the study of the conditions and policies under which reparation payments were attempted to be made after the First World War. It can be adapted to the situation at hand and presented with any desired degree of complexity. What follows is a rudimentary version of that theory.

The theory rests on the distinction between an international *financial transfer*, which is relatively painless (international "finance" being merely paper), and a *real transfer* between two countries, which represents the ultimate payment in goods and services. More specifically, the theory is concerned with determining the magnitude of the real transfer that will, under different conditions, follow any given financial transfer.

The magnitude of the real transfer will depend on a number of factors of which the most important is what is done with the financial transfer by the receiving country. The oil exporters presumably want to exchange a wasting asset for a permanent one, and in this respect they have three choices: (i) to spend the financial sum transferred to them on current imports in the effort to create, through investment, real assets at home; (ii) to hold it in the form of monetary assets; or (iii) to use it to acquire assets abroad. The analysis shall now be developed through these three basic cases, and a number of variants beginning with the simplest and most abstract ones and allowing gradually for more complexity and realism. It should be remembered however that each case represents one of the three possibilities in its pure form and that what happens in reality will be a combination of all three.

Case 1: Oil Payments Generate Imports by Oil Producers

Assume that the world consist of two countries only, an oil-exporting country *A* and an oil-importing country *B* (which may stand for all the industrial countries taken together). At the beginning their current accounts are in equilibrium, which means that each is "absorbing" (that is, consuming and investing at home) the full amount of what it produces, which in turn equals its income. Then *A* doubles or triples the price of oil, which, due to the low price-elasticity of demand in the short run, greatly increases the payment *B* has to make for its oil imports, despite some diminution in their volume. Initially the payment is made out of reserves and this represents the *financial* transfer. *A* now has a large current account surplus, *B* a deficit.

To cut through unnecessary complications, assume that the government of *A* has been running an ambitious development programme which it now greatly expands and places in *B* orders for goods and services to the full amount of its extra export proceeds. If money income in *B* is held constant, more of it is spent on oil products and less on other goods and services, which thus become available for export. When current accounts are in equilibrium again, the level of economic activity in *B* is unchanged, but its real income has been reduced by the adverse shift in the terms of trade. This reduction can best be expressed in terms of *B*'s "absorption", which can be defined as production plus imports minus exports: *B* now imports less petroleum for which it exchanges a larger part of its production than before. The reduction is equal to the amount of the real transfer, which exactly corresponds to the amount of the financial transfer.

There are oil-producing countries whose governments run large development programmes and can spend much (in principle, all) of their additional export proceeds on imports of goods and services. But there are others whose current import capacity is very limited. These will have to keep their extra proceeds in some form of assets until their domestic economic activity expands and diversifies enough to absorb in the form of current imports the whole of the savings which these assets represent.

Case 2: Oil Producers Invest in Financial Assets, Dollars or Gold

In this case three variants are considered, namely oil-producing countries investing in financial assets, dollars or gold.

Variant *A*: Financial Assets

Assume that the world now consists of four countries: the oil exporter *A* and the oil importers *E, J* and *U*. Assume further that *A* will keep all its additional proceeds in monetary assets (say, liquid reserves) and that, having no preference among foreign currencies, is willing to hold all the extra proceeds from its sales to *E* in *E*'s currency (the euro), to *J* in *J*'s currency (the jen) and to *U in U*'s currency (which, just to break the monotony, may be called the dollar).[2]

The financial transfer takes place in these three currencies. What happens next? As long as *A* continues to hold them, nothing will happen. No real transfer follows. *E, J* and *U*'s real incomes are undiminished. The higher price of oil is paid in easily printable paper.

Variant *B*: Dollars

Assume now that *A* is still willing to hold its extra proceeds in liquid reserves, but this time it has a preference for the dollar to the exclusion of both other currencies. If there were a certainty that *A* would hold these reserves indefinitely, again nothing of consequence need happen. *E* and *J* could borrow, and keep on borrowing, the dollar, essentially at the cost of printing or book-keeping operations.

But if this certainty does not exist, *E* and *J* cannot be expected passively to countenance a massive and rapidly mounting foreign debt; indeed, they must be expected to try to earn the dollars with which to pay for their petroleum. The only way to earn them is for each to run an export surplus with *U*. In this case therefore the financial transfer (in dollars) would be seen going from *E*, *J* and *U* to *A*, with the real transfer going from *E* and *J* to *U*. Three further observations on this case suggest themselves.

(a) By deciding which currency to hold, *A* can influence the direction of the real transfer to one or another country. In this particular case, *A* would be granting *U* a two-fold loan: Not only would *U* be exempt, for the time being, from the increase in the real price of oil. It would also receive the whole increased cost of *E*'s and *J*'s oil imports in the form of goods and services.

(b) Having assumed that *A* intends eventually to convert its monetary assets into imports of goods and services, its problem can be viewed as one of deciding into which country to direct the real surplus for safekeeping until that time. It would not be illogical for *A* to conclude that the largest and least trade-dependent of the three oil-importing economies offered it the best safety.

(c) The size of the real transfer from *E* and *J* to *U* would be such as to require *U*'s active cooperation for the transfer to be effected. *A*'s decision to hold dollar-denominated monetary assets would imply an inflow of dollars into *U*'s money market, reducing the interest rate. *U*'s monetary and fiscal policies would have to support this effect so as to induce a domestic boom strong enough for *E* and *J* to achieve the required surplus in their bilateral balance with *U*. Any attempt by *U* to "defend" its current account – by internal deflation, devaluation or imposition of import restrictions – would lead to a cumulative downward spiralling of trade and income among the three.

Variant *C*: Gold

(This variant is dedicated to all advocates of a return to the gold standard.) While *A* can control its terms of trade by pegging the oil price to the export price index of the oil-importing countries, the latter could still be tempted, after a period of price stability, to escape through inflation their real payment obligation on the monetary assets which *A* had accumulated in the meantime. This global inflationary risk would be banned of course under a monetary system which would be based on gold. In such a system however it would seem logical for the security-seeking and mistrustful *A* to wish to hold all its reserves in gold – which would be practically the same thing as fixing the price of oil in gold. With low price-elasticity and high income-elasticity of demand for oil, central banks of *E*, *J* and *U* would soon be emptied of gold; and with sharply increasing costs in gold mining, the world price of gold, and thus also that of oil, would soar. The gold-based monetary system would thus end. To the extent that gold can be considered a "financial" instrument, the "financial" transfer would be taking place from *E*, *J* and *U* to *A*, the real one from *E*, *J* and *U* to the gold producers.

In view of the various risks connected with monetary assets in the world of unstable exchange rates and uncontrollable government budgets, it is unlikely that risk-avoiding *A* would be willing to accumulate its extra export receipts in this form for long. It would be quick to see the advantages present in holding real assets.

Case 3: Oil Producers Invest in Real Assets either Worldwide or in One Country

Thus the analysis arrives at the third case in which the oil producers invest in real assets either worldwide or in only one country.

Variant *A*: Worldwide

To spare its customers *E*, *J* and *U* unnecessary balance-of-payments difficulties, *A* would agree to invest in real estate and industry in each importing country all the extra proceeds of its sales to that country.[3] No real transfer would follow and current income in *E*, *J* and *U* would remain undiminished. Each of the three would have a deficit on current account, offset by an increased inflow of long-term capital. The purely financial transfer, however, from *E*, *J* and *U* to *A* would be followed by an equivalent transfer of *real property titles* in the same direction.

Variant *B*: One country

Here *A* decides to invest all its extra proceeds in real assets in only one of the importing countries. This brings the discussion back to Case 2*B*. The sole new possibility to be noted in this case is that, in the process of gradually channelling its proceeds from worldwide sales into one country, *A* would be more likely than in the preceding variant to acquire a growing share of the total real wealth of the country of its investment choice.

Many more variants are thinkable but they would change nothing essential. Cases 2*B* and 3*B* could be resolved either by a large current account deficit of *U* or by a large long-term capital outflow from *U* to *E* and *J*. In this case there would be a transfer of real assets (property titles) from *E* and *J* to *U*. In other words *U* would again be borrowing short and lending long as was its wont to do in the period of the dollar glut, except this time on an approximately ten times larger scale.

CONCLUSIONS FROM THE THEORY

What conclusions can be drawn? If *A* could spend all its extra proceeds on its own development, *E*, *J* and *U* would have a real transfer problem, but not an unsolvable one. They would feel no deflationary pressure, since export demand would "take up" those resources on which spending declined as oil products and energy in general attracted a higher proportion of domestic expenditure. The once-and-for-all increase in the oil price would represent a real transfer of 1–2 per cent of their gross national product (GNP) (that is about one-half of their annual GNP growth capacity), and, after the inevitable structural adjustment, growth capacity of these economies would remain unaffected. It is indeed probable that the growth of global real income in this four-country world, after an initial deceleration due to the adjustment friction, would be stimulated by the international income redistribution implied by the transfer and be faster than before.

A more serious difficulty stems from *A*'s inability to spend all of its additional proceeds on current imports. In itself this again need not have any deflationary consequences: on the basis of *A*'s lending to or investment in the importing economies, the latter can generate the additional expenditure to "take up" those resources from which domestic spending shifted to oil products. The difficulty arises rather from the fact that *A*'s asset-holding decisions are likely to create balance-of-payments problems among *E*, *J* and *U*. One or more of them could be forced into borrowing so large that it would not be politically acceptable as a long or medium-term proposition, be it borrowing from an offshore currency market, another country or an international institution. (*A* itself, holding a large amount of monetary assets and fearing inflation, would be likely to protest against large-scale lending by international institutions.) Similarly, *E*, *J* and *U* are likely to find it objectionable to exchange, period after period, portions of their accumulated wealth for a necessary current input. They cannot see themselves "living off capital" in this way, besides being aware and wary of the political control that foreign ownership of wealth located within their national borders would confer. But as a group they have no way of earning an additional, say, $40,000m by running surpluses with each other; and they know that if they tried it they would ruin their economies.

So what can they do? There seems to be only one answer: to cooperate with *A* in using that part of its export proceeds which it cannot spend on current imports for the creation of *new* assets attractive enough for *A* to hold.

As for new real assets, the question arises as to where they should best be created. Let us now enlarge the model into a five-country world by introducing a poor oil-importing country *D*, whose prospects of economic development have been gravely jeopardised by the oil price increase. If *A* devoted its "unspendable surplus" to new investment in D, it would more than offset the additional amount that *D* would have to pay for oil. *D*'s chance of further development would thus not only be safeguarded. It would be enhanced. And since the equipment, machinery, engineering and management know-how for this new massive investment would have to be imported from countries other than *A*, the three industrialised oil importers, *E*, *J* and *U*, would have an opportunity to earn their oil-dollars without cutting each other's throats.

But there is one condition. This investment could create industrial assets worth holding only if the projects were conceived and planned in a framework of an open world economy. Recent decades abound in evidence that sound industrialisation is impossible on the basis of inward-looking strategies in which industries are planned to produce for a delimited market, whether large or small. Even in the largest of the national markets, in the area called D, the benefit forgone in renouncing international production specialisation would reduce the return on capital and *A*, carefully comparing yields on alternative investment opportunities, would probably invest elsewhere. In other words a significant part of the output from the new investments in *D* would have to be expected to enter the export market. This in turn might require some rethinking on the part of *E*, *J* and *U* of their commercial policies.

NATURE OF THE OIL PAYMENTS PROBLEM

We turn next to the second and more empirical aspect: the possible extent and the exact nature of the payments problem. In 1973 world exports of crude oil were close to 11,000m barrels. How would demand react to the new price levels? This reaction was seen as being determined by three demand-elasticities:

(a) with respect to price (although the elasticity may be low, the price increase was large);

(b) with respect to income (this elasticity may be higher than the preceding one, but income was unlikely to increase very much in 1974 and it could even have declined); and

(c) with respect to substitution (there is always some possibility of producing domestic substitutes for an import that has risen in price).

Making only moderate assumptions about those elasticities, it was expected that demand for oil imports in 1974 would run some 5–7 per cent below the 1973 level. At the new price[4] the anticipated 1974 demand for 10,000m barrels represented a revenue of $85,000m for the oil exporters, of which some $68,000m would be paid by industrial and some $17,000m by less developed importing countries. The expected increase in the oil bill of all importers combined was some $65,000m from 1973 to 1974.

What part of this revenue could the oil exporters spend on current account? Both the trade and balance-of-payments statistics of the oil-exporting countries are not very reliable and are late. The most recent set of statistics at the time of writing is for 1972.

One has to guess therefore how much imports increased in 1973, and on that guess base a more hazardous one, as to the increase in 1974, when export revenues have more than trebled. A conservative estimate for the increase in the volume of merchandise imports would be 20 per cent in 1973 and 25 per cent in 1974.

Data on service payments and transfers[5] are always estimates, less reliable than straight import statistics. What they indicate however accords with expectations. Primitive but rich economies deriving their revenue from a single resource should have a much higher ratio of service to merchandise imports payments than the more diversified economies.[6] Consider conditions in the Persian Gulf countries: large numbers of immigrant workers; foreign construction firms building ports and factories which will operate with imported management, skilled labour and proprietary knowledge; health services manned by foreign staffs and foreign mercenaries in the armies; travel abroad, financial aid to other countries; and transfers to Arab religious and political organisations. Payments for arms, not usually shown in trade statistics, should also be included in the service account.

Furthermore, if the revenue from the main export suddenly increases, and if a good part of the increase is put into domestic development, one would expect the service payments of such an economy to increase more rapidly than its payments for merchandise imports. Let it be assumed that for all oil exporters combined, service payments represent 40 per cent of merchandise imports payments and that they increased 30 per cent in 1973 and 35 per cent in 1974. This too may be a conservative estimate. On these assumptions oil exporters were expected to spend some \$35,000m on current account in 1974. The collective "petroleum" deficit of the importing countries would have been \$50,000m.

How will the problem develop? This depends essentially on two factors. How will the demand for oil imports, and the "import-absorption capacity" of oil exporters, change in the years to come? Assuming the 1974 price of oil to remain unchanged, future demand for oil imports will depend mainly on (i) GNP growth in the importing economies and (ii) the progress they can make in energy saving and in the development of domestic energy sources. It can be taken for granted that even if the importing countries recover their full-employment rates of growth in 1975 or 1976, oil imports will grow considerably less rapidly than in the past, as the pattern of future growth is bound to be more energy-saving at the new relative price for energy. Taking account, in addition, of domestic substitution possibilities, it is not certain that oil imports will grow at all.

The "real" demand for imports is a function of economic development, the main determinant of both being the organisational ability of the society in question. The import-absorption capacity even of the desert economies will grow as their increased income enables them to buy "organisation" abroad; and its growth can be accelerated further by appropriate cooperative policies of the oil-importing countries. If organisational ability is the main limiting factor, the institution of corporate joint-ventures, in which oil capital would be combined with engineering and organisational know-how of firms from industrial countries, would appear to be the most effective way of increasing the import demand of oil-producing areas. This of course is what the technology-transfer agreements offered by the oil-importing to the exporting countries are all about. Note also that, apart from the question of security of supply which these agreements seek to promote (but which is considered less important now, while the payments problem has come to dominance), it makes no difference to the oil-importing industrial countries whether such joint-ventures are located in Kuwait or in India. Indeed labour supply may be the limiting factor to industrial joint-ventures in the Arab oil-producing countries, whereas a combination of oil capital, Western engineering and organisation, with labour of other developing countries does not face any *a priori* constraint.

Assuming oil imports to grow by about 2 per cent per year, retaining the rates of increase of current account expenditure of oil exporters used in the estimate for 1974, and calculating a 10 per cent yield for the capital surplus which the oil exporters will cumulate abroad, the current spending of these countries would grow to equal their current revenue in five to six years. The accumulated surplus would in that time reach some $250,000m. If the absorption capacity and the demand for oil imports continued to grow at the assumed rates, the whole accumulation of their capital abroad could be reabsorbed by oil-exporting countries in four more years.

For a group of countries whose ambitious economic development programmes are not inhibited by a lack of foreign exchange, it should not be a problem to sustain an average growth of real imports at 25 per cent per annum. The above calculation can thus be considered to refer to a situation in which the real price of petroleum is stable (for example oil is indexed to the price level of exports of manufactures). If, on the other hand, only the nominal price of oil remained stable, its real cost being reduced by continuing inflation abroad, the calculation should be adjusted by adding the rate of inflation to the rate of increase in real absorption capacity. Both the duration of the payments problem, and the magnitude of the accumulation of oil capital abroad, would be substantially reduced in this alternative.

Distribution of the Oil Deficit

The size and duration of oil payments, borrowing and debt accumulation evidently enter the payments problem, but are not of its essence. I have already shown that no payments problem could arise in a world consisting of one exporting and one importing country, whatever the size of the payment. The problem, with its implied danger of cumulative deflation, arises from there being many oil-importing countries, each with a different balance of external payments and indebtedness, a different degree of dependence on oil imports, and different prospects for energy independence; and it consists in distributing the collective deficit among these countries in a way that is "acceptable" to all.

A few probably could pay the whole increase in their oil bill in goods and services. Few if any will presumably wish to do so, for it would necessitate a significant cutback in current consumption and investment; and hence less growth. Few if any, on the other hand, will be willing and able to borrow the whole amount. But between these two extremes, each country will have a preference as to the proportion of its oil payment to be made in real deliveries and in debt. The problem, in other words, is to reconcile these preferences so that countries with a pronounced "cash-and-carry" preference do not force the others into more borrowing than they feel they can afford.

In consequence of the exceptionally strong boom of 1972–73, several countries had been running large deficits and experiencing some difficulty in financing them already before the price of oil was raised. Under normal conditions they would restore their current balance gradually by letting domestic demand grow less rapidly than the GNP. With a new large deficit superimposed on the old one, with their lines of credit heavily drawn upon, and with economic growth slackening generally, they may be forced to reduce domestic "absorption" absolutely. This would mean less imports from, and/or more exports to oil-exporting *as well as* other importing countries; and if some of the latter (for example those in the next weakest balance-of-payments position) attempted to defend their current accounts in turn, a cumulative deflationary process would be set in motion. Clearly if the level of real expenditure in the oil-importing countries as a group is not to decline, the domestic demand (and possibly import) restrictions which countries in the weakest current-account position cannot avoid imposing must be compensated by a stimulation of domestic demand in countries which were in surplus before the onset of the new oil payments and a consequent drawing down of their current-account balance.

There is an element of paradox in this interpretation of the payments problem. The collective deficit of oil importers, with the distribution of which we are concerned, is an unambiguous and measurable concept. In the case of any one importing country, on the other hand, an "oil deficit" is a notion without operational content.[7] The distribution of the oil deficit in other words can be followed only in terms of overall changes on national current accounts.

At present the pattern of these changes cannot be foreseen except in the broadest outline. Oil-importing developing countries cannot in general be expected significantly to increase their sales to oil exporters, and only a few of them are in a position to borrow on commercial conditions to pay for oil imports. Their additional oil payment needs, net of any direct assistance they can receive from oil exporters, must therefore be met either (i) from additional assistance given by the developed countries or (ii) from foreign exchange earned or saved in trade. Each of these alternatives implies a reduction in the current-account surplus which the developed countries as a group normally run with the developing ones. A change is also foreseeable in the pattern of current-account positions among the developed economies. In the first quarter of 1974 the current accounts of the three largest ones combined were in surplus. Should this position be maintained, it would be obviously difficult to devise a generally acceptable way of distributing a collective deficit of some $40–50,000m among the twenty-odd smaller economies of the Organisation for Economic Cooperation and Development (OECD).

On certain assumptions, our interpretation of the payments problem can be shown to be identical with the more widely debated one which emphasises the need for redistributing (recycling) the credit implied in the oil producers' decision to export more petroleum than is necessary to cover their current import demand. To concentrate however on providing the finance for "such deficits as may arise" presupposes an effective and largely automatic balance-of-payments adjustment mechanism. Since the mechanism is currently undergoing repairs and its present effectiveness is uncertain, an interpretation of the payments problem which concentrates on the current account has the advantage of pointing to the more urgent need for a close and practically continuous coordination of national demand management policies.

INTERNAL ADJUSTMENT

In the short run the oil-importing countries face collectively the problem of avoiding deflation. In the long run each of them faces the problem of living with and eventually repaying a large external debt. To live well with debt requires considerable skill in using one's options, of which there always is a range.

On Living with Debt

The notion that the increase in oil prices is a tax on oil users is useful in exploring the range of options open to oil-importing countries. The revenue of this tax is collected by the oil companies in domestic currency. It represents a claim to an amount of domestic goods and services on which this money would have been spent had the oil price not risen. To transfer this amount to oil-exporting countries the companies will, however, need foreign exchange. Assume this amount of foreign exchange to be borrowed by the country's central bank from offshore currency markets and sold to the oil companies for domestic currency. Both amounts represent claims to resources: the revenue in domestic currency a claim to domestic resources, its foreign exchange equivalent a claim to resources anywhere.

A part of the country's oil bill will be paid in additional exports and an equivalent amount of foreign exchange will return to the central bank as a credit against its debt to the offshore currency market. The equivalent in domestic currency of the net oil indebtedness abroad corresponds to the amount of domestic resources that are still waiting for redeployment. How well the country will live with its growing external indebtedness depends on the use to which these resources will be put.

Inside the economy the potential demand for them is strong. Consumers have just suffered a cut in real income as oil products and energy in general have risen in price; if the government reduced taxes the temporarily unemployed resources would be quickly taken up again in a return to former real consumption levels. At the same time the increased price of oil sets up incentives for new investment. It is important for the subsequent argument to have a correct appreciation of the nature and extent of these incentives.

Between 1950 and 1970 the real price of crude oil[8] declined by more than one half. The market mechanism reacted with a high degree of efficiency. The instances in which the steadily cheapening energy was substituted for more expensive resources are infinite. It would be difficult to overestimate this substitution as a driving force of economic development in that period. Aluminium, a highly energy-intensive metal, was increasingly substituted for steel and other metals which are more labour-intensive. The capital cost of housing construction was reduced by building a thin shell of a house and regulating inside temperature by a conditioning "energy input". The development of large "self-service" shopping centres at the outskirts of cities can be interpreted as a substitution of cheaper automobile transport for more expensive sales services in the city centres. The technology of land, sea and air transport was decisively influenced by the availability of cheap oil. Even the late Green Revolution amounted, in the last analysis, to an effort to compensate for the natural disadvantages of basic-food farming outside the temperate zone by increased energy inputs in the form of petro-chemicals and straight energy for water-pumping. These few examples indicate the pervasive nature of this substitution. There is no reason to doubt the market mechanism's ability to act equally efficiently in the opposite direction, finding cheaper substitutes for the one product that has suddenly become very much more expensive; and all the new substitutions will require new investment.[9]

A firm borrowing a large amount must convince the lender that the purpose to which the money will be put will sufficiently increase its cash flow to allow it to establish a "sinking fund" in which the amount borrowed will be accumulated by the time the loan falls due. A borrowing government has to behave in a similar way, not necessarily because it is so required by the lender, but because it wants to avoid the need for sharply reducing domestic consumption and investment at the time when the loans come due. Its "sinking fund" (an accounting fiction in both cases) would consist of productive assets enhancing the flow of goods and services that will ultimately be demanded in repayment of its borrowing, or alternatively, that would otherwise have to be imported (for example oil). The surplus revenue which the oil exporters temporarily cannot spend through their current account would in this way be invested in new productive enterprise in the importing countries.

Here however it is necessary to restate the main point of the preceding analysis. It is that resources are made available for redeployment, and new investment incentives set up, by a (relative) increase in the oil product prices charged to consumers. A realistic analysis has to recognise that in this respect the governments of the importing countries face a dilemma. Since their taxation of oil products has been heavy, they can by reducing these taxes determine the rate at which product prices to the consumer will rise to reflect the full increase in the price of crude. And the more sharply these prices rise, the more severe will be the "adjustment friction", a notion comprising not merely the extent of the reallocation of resources dictated by the price change, but also the additional loss of output due to the social struggle through which the cut in real income is apportioned among the various interest groups in the importing economies.[10]

On this background, the basic options facing the governments of oil-importing countries can be presented in the form of a few starkly simplified cases. By reducing or eliminating excise taxation of oil products a government could fully offset the increase in company prices. With relative prices constant there would be no change in the pattern of final consumption, no new investment incentives and no adjustment friction. The deflationary effect of the current-account deficit would be offset by the deficit in the government's budget which would return into circulation the amounts of domestic currency sold by the oil companies to the central bank for foreign exchange. The foreign borrowing would in this case be used to finance consumption.

If the government stopped the budget deficit by raising general taxation, an amount of domestic resources corresponding to the increased cost of oil would be released from domestic spending. A reduction in the interest rate would stimulate investment demand for these resources. The growth resulting from the investment however would be of the old, excessively energy-intensive pattern. In both these cases oil imports would continue to grow at the high rate established in the past, and foregn indebtedness would be rising rapidly.

Letting internal oil prices rise by the full amount of the external price increase would, on the other hand, set up the correct incentives for a new growth pattern. In this situation the decision how to use the resources which constitute the real domestic counterpart to foreign borrowing – that is, whether to release them for consumption (by reducing taxes) or whether to invest them (by reducing the rate of interest) – is essentially a decision as to the speed with which this new pattern will be realised. Most of these resources were employed in consumer industries; consequently to use all of them for investment would imply an extensive resource transfer combined with considerable friction. A tax reduction which would mitigate the effect of the price increase on real disposable income would reduce the amplitude of the transfer, but by the same token extend the period of transition. It could not be more than a temporary palliative. For ultimately, as external borrowing continued, all the counterpart resources would have to be invested.

On the Need for Structural Adjustment

Countries with no domestic oil cannot but allow consumer prices of oil products to rise because (i) they have to check the growth of consumption and import dependence and (ii) after some initial "recycling" their external credit will be only as good as the investment projects they have ready to be financed by it.[11] The need to allow domestic prices to rise is less obvious in the case of countries that do not import oil.

Imagine a fairly large industrial economy which is self-sufficient in energy. As the external oil prices quadruple, numerous arguments will be made inside and outside the government for an export embargo which would prevent domestic prices from rising at all. Two will be used most persistently:

(a) The extensive change which the oil price increase

would dictate in the expenditure pattern would imply high frictional (structural but temporary) unemployment. The cost of that unemployment, measured in terms of output forgone, would be considerable. Would it not be arbitrary to impose this cost on a society in which nothing changed since it does not have to import oil?

(b) The development of alternative energy sources is highly capital-intensive, and energy-saving also requires additional investment. Forcing the economy to react to the external oil price increase would thus imply the diversion of large amounts of capital from a (potential) expansion of consumer-goods into producer-goods industries, which would be inflationary in the short run and reduce the growth of consumption and material welfare.

Obviously what is a correct policy can only be determined with respect to future price behaviour. Should the external price increase be only temporary, internal adjustment beyond a certain limit (set by the importance which the society placed on the security of its energy supply in the long run) would be wasteful. This is the explanation as well as justification of the temporising manifest in all capitals in 1974. One does not embark on important changes in the structure of the economy before a reasonable view of the future can crystallise from evolving events, and that takes time. This chapter however is not engaged in an estimate of the empirical probabilities of the international oil price declining again, staying where it is or rising further in the future. The purpose is merely to examine certain arguments on the hypothetical assumption that the international economy has moved *permanently* into a phase of high-cost energy.

With economies interdependent in a global system, the intrinsic value of any product is determined in the world market. The intrinsic value of oil reserves even of a country which does not have to import oil has been raised by the world market price rise. Trying to "protect" that country from an external price increase on a product in which it does not have to trade would amount to keeping the private user cost of that product below the social cost (international price). There would be several elements of this excess of social over private cost.

The oil reserves of the country under discussion are finite. Refusing to change the domestic price, the country would not only be prolonging a pattern and a rate of consumption that are now clearly wasteful;[12] it would in fact be increasing the drain on domestic reserves. The discrepancy between the domestic and the world market price of oil would give a great advantage over international competition to those of the country's industries which were intensive users of oil-based inputs such as energy (aluminium industry) or naphtha (petro-chemicals). These industries would tend rapidly to increase their production and exports.[13]

Secondly, the country would approach the depletion of its reserves with a more oil-intensive pattern of production than would have been the case if the adjustment of the domestic to the international price had begun earlier. The capital cost (in the broadest sense) of adjusting the more thoroughly distorted industrial structure would be correspondingly higher.

It is important to understand the nature of the distortion. While for example export industries producing basic materials such as aluminium and petro-chemicals would expand, industries closer to the finishing stage would see their exports gradually shrink. Their products, whether producers' equipment or consumers' manufactures, developed in and produced for a cheap-energy economy, would be increasingly ill-adapted to foreign demand already molded by the scarcity of energy. With a growing part of these industries' output saleable only in the domestic market,[14] the productivity of their existing apparatus would suffer. Even more far-reaching would be the effect on technology, which has been the major source of economic growth in the modern era. Technological innovation is responsive to relative scarcities of individual factors or inputs, and it has been so rapid in the past because the research and development effort of each society could draw on similar work, pursued in the same direction, in all the others. By refusing to adjust its domestic oil price, the country would be dissociating itself from technological research abroad which the energy shortage was impelling in a wholly new direction. These, then, would be the current social costs of staying with the old oil price; it would not be too exaggerated to call them the costs of gradually opting out of the international division of labour. The capital costs of the postponed readjustment of the industrial structure would amount in effect to a loss of a significant proportion of the savings embodied in capital equipment in existence when the price of domestic oil finally had to increase.

Alternative Modes of Adjustment

Adjustment of economic structures to a large increase in the relative price of so important an input as oil, which effectively governs the price of all forms of energy, inevitably implies some unemployment. Demand for energy-intensive products and products complementary to them declines, while demand for substitute products rises. Production factors are specialised and localised; even the investment process is specialised in the sense that a radical change in the investment pattern creates organisational problems the solution of which takes time. Before new jobs are created in the industries on which the new pattern of demand converges, and before local labour is trained for them or adequately skilled labour moves to them, workers of industries from which demand ebbs away and new entrants into the labour force, will be unemployed. The unemployment rate may rise quite sharply.

There is a danger of cumulation. If the unemployed of the adversely affected industries were to become destitute, or – a more likely danger – if the rest of the society, suddenly uncertain about the future, increased its saving sharply, products which would otherwise still be wanted would not get produced and the frictional unemployment would be accompanied by unemployment due to a deficiency of aggregate demand. Unemployment of the latter type can however be controlled by monetary–fiscal policy[15] and is not at discussion here.

But it is only the latter type of unemployment, occurring without a significant change in the prevailing tastes and wants, whose cost can adequately be measured in terms of the output forgone.[16] The notion of the social loss due to frictional unemployment is a much more subtle one. It cannot be measured by the output that the unemployed workers could have produced if they had stayed in their jobs. For how much could they produce and, more importantly, how should that output be valued? When, in consequence of a large increase in the price of petrol, the demand for new eight-cylinder automobiles declines, the value of such a car does not sink to zero. But consider a firm specialising in the production of such a car, its equipment unadaptable to other functions.

As the petrol price rises the firm's demand curve shifts leftward. If the firm were to produce, at the new lower price, enough cars to keep its labour force fully employed, it would suffer a loss. Its plant and equipment would not be earning enough to replace themselves. If the new full-employment price for the firm was sufficient to cover the average variable cost of producing the car, either at the existing wage rate or at such lower wage rate as the workers would accept in order to continue production, it would contribute to the welfare of the society to go on producing that car until the first vital piece of the non-replaceable equipment wore out. This is in fact the point at which a firm in a competitive market would give up production. The notion of "output forgone through unemployment" can only refer to output that was lost unnecessarily, for example, through a failure of fiscal–monetary policy. Production abandoned because tastes changed, and the price at which consumers would be willing to go on buying the product would not cover the full cost of producing it, cannot be considered "forgone" in this sense.

Now we are ready to broach the difficult question whether and on what grounds a gradual adjustment of the domestic price of oil to the new international price is to be preferred to an immediate full adjustment. The temper of the times is such that to the majority of people the word "gradual" suggests, almost regardless of context, a process preferable to that suggested by the words "sudden", "abrupt" or "speedy". With respect to industrial adjustment in general, reputable economists are on record as preferring gradualism.[17] Yet is there a rational justification for the preference? Will a gradual adjustment be accompanied by a smaller amount of frictional unemployment than a quick adjustment?

Basically the government of our self-sufficient country has three adjustment options. It can by a number of different means (including a free export policy) let the domestic oil price join the international price in one step. It can announce that the domestic oil price will be brought up to the international level in, say, four equal and equidistant steps over the next four years, and proceed to do so. Or it can plan and do exactly that without announcing its plan.[18]

The theoretical model of perfect competition achieves its efficient, instantaneous adjustments by virtue of, among other things, the perfect foresight which it postulates for all firms and individuals. The foresight required by this model amounts to omniscience. Let us choose a somewhat lower level of abstraction, an economy with a "reasonable" degree of both competition and foresight, in which, however, certain variables are controlled by government policy, and investment is a process in time. Firms and individuals are exercising "reasonable foresight" if they attempt to anticipate future developments allowing for varying degrees of uncertainty.

It is easy to see that in such a society there would be little difference, in the time-profile and cost of adjustment, between the first and second option. If the gestation of an average investment project were four years or more, there would be no difference between the two adjustment processes as to the level of investment activity. Some investments can be brought to fruition in less than four years, and to the extent that they would be profitable only at the international level of energy prices they would be postponed to a later phase of the adjustment period.

Neither would there be much difference with respect to the rate at which the economy would reduce or abandon the production of energy-intensive goods. To return to the earlier example, the present value of an eight-cylinder automobile acquired after the domestic petrol price has fully adjusted to the international price must be less than the value of the same car bought with the prospect of the domestic petrol price rising gradually to the international level over a four-year period. Therefore more of these automobiles would be bought by, and contribute to the welfare of, rational consumers deciding on the basis of what was known about the future. The relatively larger production would imply less frictional unemployment in the period of gradual adjustment. As against that however one would have to count the increased consumption of petrol at a socially subsidised price. The social welfare gain from the gradual but foreseen as against the immediate oil price adjustment would thus be equal to the value of the additional cars produced in the former course, diminished by the loss of economic welfare associated with the "subsidy" (difference between the domestic and international petrol price) to their petrol consumption in the first four years of their life. This would almost certainly be a positive amount, but how significant could one expect its order of magnitude to be?

The welfare gain would depend on there being no miscalculations and subsequent disappointments by the buyers. A rational buyer would of course capitalise the net value of the future services of the car by his personal rate of time-discount. Being rational – that is, in the present context, aware of the current state of the science of economics – he would however have a problem as to the pricing of the future services to be discounted to the present. If he valued them at the prices of the moment[19] he would be bound to overvalue his acquisition and to come to grief. We must hope for him to be rational enough to realise that, the present forms and systems of transportation being highly wasteful of energy, it will be a close-to-the-top priority of the adjustment process to develop more convenient and pleasant mass-transit systems than we have now, an improvement which will diminish the relative value of the services of his private car. One must also hope for him not to be hoping to be buying prestige with an eight-cylinder car, a hope to which the current system of social values lends considerable support. A fully rational buyer of such a car would take into account the possibility that a few years hence, well within the useful life of the car, his passage in the contraption might be acknowledged by a shrug: "Some nut". Given this degree of foresight, there could not be much difference in the number of eight-cylinder cars produced in the process of immediate as against gradual adjustment. The "social welfare" gained from producing more such cars because of a failure of foresight on the part of the buyers would be offset by their subsequent disappointment: no net gain in utility.

The main difference is between the first two adjustment options on the one hand and the third on the other. Those who would prefer the third approach would have to be counting on an effect comparable to the "money illusion" of monetary theory. At each raising of the domestic oil price a number of firms and individuals would believe it would be the last; and from that the advocates of this policy would have to be deducing that the decline in the production of eight-cylinder cars would be less steep than under the preceding alternatives and, consequently, there would be less frictional unemployment. In fact, though, it is imposible to see how this adjustment course could be preferable to the other two, how less information could lead to a better social result than more information. With the increased degree of uncertainty, the number of errors and miscalculations would be bound to increase. There would be less investment in "appropriate" projects than under the other two approaches and there would be a number of investment projects which each subsequent (unanticipated) domestic oil price increase would make uneconomical. At each price increase there would therefore be more redundant capacity in some, and more capacity shortages in other industries than under the alternative approaches. New entrants into the labour force would acquire an inappropriate mix of skills, which would ensure a higher level of structural unemployment in the future. The economy would be turning out many goods bound to disappoint the customers' expectations. Last but most important, it would be wholly unrealistic to expect that firms and consumers, left in uncertainty about future prices, would maintain the level of their expenditure. Future analyses of the 1974–75 recession may indeed show that its severity was due to the unusually high degree of uncertainty which the international oil price increase imposed on all industrial societies.

CONCLUDING OBSERVATIONS

In the last analysis, the industrial countries will be paying for oil by their unique and most valuable resource – their technology and organisational ability – which is required to put to productive use the increased global saving brought about by the oil-exporters' inability to spend all their current revenue. The importing countries even have considerable freedom of decision as to where the new productive structures resulting from this combination are to be allocated: whether in the oil-producing countries, in the developing countries or in their own countries.

The internal adjustment of the industrial countries to the high price of oil will require very large amounts of capital. The amounts borrowed from the oil exporters will in any case have to be supplemented by additional domestic saving. The extent of this need will be strongly influenced by government programmes aiming at energy independence. The more ambitious these plans, the less capital will be available for economic assistance to developing countries.

The social welfare loss from frictional unemployment cannot be measured by what the unemployed would have produced in the jobs they lost, but only by what they could have produced in their new jobs *had they found them earlier*. This loss is the difference between the best a society could do in reallocating resources, given its existing organisational means, and what it actually does. If the society thinks it is doing its best, there is no loss.

There are serious economic arguments for believing that the rate of exploitation of non-replaceable natural resources dictated by the market rate of discount is unduly biased against future generations. A rational society thus might consider a government action to reduce what might be an excessively high rate of exploitation. If the new international level of the price of crude oil were expected not to decline, government policies intended to make the adjustment gradual would, if they were successful, in contrast raise the rate of exploitation beyond what the play of market prices would dictate. But responsibility to future generations does not end at leaving some resources in the ground. A greater and more immediate responsibility consists in leaving to the next generation an appropriate, useful structure of productive assets and technology. The force of this argument may be better appreciated if the reader realises that his own retirement income depends on it.

NOTES AND REFERENCES

1. This chapter began as a paper presented at a meeting of the Trade Policy Research Centre in London, in January 1974, which was subsequently published as Jan Tumlir, "How the West Can Pay the New Arab Oil Bill", *Sunday Times*, London, 3 February 1974. The analysis was developed further in the spring of 1974 (Nature of the Oil Payments Problem and On Living with Debt), and published as "Oil Payments and Oil Debt in the World Economy", *Lloyds Bank Review*, London, July 1974. That version has been extended for the present volume. It has to be stressed that the views expressed here are the author's and are not to be interpreted as reflecting those of the Secretariat of the General Agreement on Tariffs and Trade (GATT) in Geneva.
2. With no exchange transactions taking place, this variant could also be presented as a case of two countries only.
3. This variant could again be presented as one involving two countries only.
4. The "price" considered here is not what the importing countries pay to oil companies but what the exporting countries retain. The $8.50 per barrel used in the calculation is a "guestimate" averaging the producing governments' "take" from company oil, resale price of participation oil and future auction prices of royalty oil.
5. Having assumed a "net" price for exported oil, "invisible imports" of oil-exporting countries are taken net of the income of oil companies.
6. The demand of oil-producing countries for foreign services would be a worthy and interesting subject for research.
7. How should the additional exports made, or imports forgone, in response to the oil payment be taken into account? Should the attempt to identify an

"oil deficit" be limited to the current account, or should oil-related capital flows be considered as well; and how should the latter be distinguished from other capital flows? The increase in the oil bill is the only measurable change; but note that a government which would agree not to act on the current account until it deteriorated by the full amount of this change would in fact be agreeing not to act until after the worst had happened.

8. Average c.i.f. price of crude divided by implicit GDP price deflator of the United States. Between 1950 and 1970 this index declined by 58 per cent.
9. See Chapter 7 below on energy-saving courses of action.
10. A pure example of this social friction was the reaction to the oil price increase of the truckers in the United States; the same factor was an element in the British coal-miners' strike at the turn of 1973–74.
11. In these countries one often hears a rather peculiar argument against adjustment. If we adjust, it runs, and commit ourselves to expensive alternative energies and the others do not, and if the international oil price comes down again, our industries will be at a competitive disadvantage internationally. Mercantilist thinking is, almost by definition, fixed-exchange-rate thinking. Since energy cost is present in all production, and since in all except a very small number of industries it is a very small proportion of total costs of production, the "disadvantage" would be cancelled by a small movement in the exchange rate. Besides even if the oil price came down, which government would wish to return to the pre-1973 pattern and growth of oil consumption?
12. They were satisfactory or economical only as long as it could have been assumed that imports would be available at the same price when domestic production declined.
13. It is not easy to foresee the reaction of the competing and the importing countries. In the theory of international trade, restrictions placed on the export of raw material which reduce the domestic price of that material below its international level, while export of products from that material is free, can be shown to be equivalent to subsidising the export of the processed product. There might therefore be protests and defensive actions. On the other hand, countries desiring to reduce their own consumption of energy would not care too much to continue being exporters of energy-intensive products, and might be even willing to import them in increased quantities, maintaining only such domestic production capacity as needed for minimum insurance against disruption of external supply.
14. And possibly in the markets of some of the oil-exporting countries.
15. The practical difficulty of distinguishing frictional from cyclical unemployment in time for appropriate measures to be taken makes that control less than perfect. This uncertainty also explains the temporising by policy-makers in 1974, especially if it is recalled that the international oil price increase occurred in a situation when an exceptionally strong and generalised boom had already passed its peak and normal cyclical recessions could be foreseen in most of the major industrial economies.
16. The adequacy of this measurement is not above dispute. There is not merely the "doing without" desirable goods, it is said; there is also the satisfaction forgone in producing them, measured by the humiliation of living on the dole (net of the welfare value of the additional leisure), and

also the sympathetic suffering of the society at large (which includes the fear that many employed must have of losing their jobs, too). While the politician perceives these "discomforts of not producing" as separate from the value of the output forgone, the economist suspects that some double-counting may be involved here.

17. For example it is stated in Richard N. Cooper, "Liquidity and Payments Adjustment", in *International Reserves* (Washington: International Monetary Fund, 1970) pp. 127–28, that "the rationale for reserves to finance imbalances arises from the real cost associated with adjustment or, when adjustment is unavoidable, from speedy adjustment".
18. I owe this idea of three basic options to my colleague Richard Blackhurst at the GATT Secretariat in Geneva.
19. This is what Sir John Hicks, "Capital Controversies: Ancient and Modern", *American Economic Review*, Papers and Proceedings, May 1974, p. 308, seems to be recommending as a proper way of arriving at the true values of capital goods.

CHAPTER 5

Trade Prospects for Developing Countries after the Rise in Oil Prices

ALASDAIR MACBEAN

In Chapter 4, Jan Tumlir drew attention to the transfer problem that will develop as a result of vast surplus revenues accumulating in oil-exporting countries. He suggested that the problem might be overcome through the investment of petro-dollars in developing countries, pointing out that it is unlikely to take place though if the products of that investment are excluded from the markets of developed countries. What is called for is a reform of commercial and industrial policies in the industrialised world to effect a shift from developed to developing countries of manufacturing activities in which the latter have a comparative advantage by virtue of large reservoirs of low-cost labour. In Chapter 9, George Ray urges that developed countries should move out of those industries which in order to compete have become, under the influence of heavy protection, very capital-intensive, so that resources can be released for technologically more advanced industries.[1]

It is not the purpose of this chapter to pursue these ideas. Instead the aim is to assess the prospects for the trade of developing countries against the background of *inter alia* higher oil prices.[2]

TRADE OF DEVELOPING COUNTRIES

It has never made a great deal of sense to lump together the many and diverse nations of the Third World as if their problems and prospects were largely identical. After the commodity boom of 1972–73, and the subsequent quadrupling of oil prices between 1973 and 1974, it makes even less sense. The first step in any analysis of the history and the prospects for the exports and imports of developing countries is to adopt a grouping of these nations which makes at least some generalisations tenable.

One self-evident group is made up of the members of the Organisation of Petroleum Exporting Countries (OPEC). Of course Nigeria and Indonesia are very different from Kuwait, Abu Dhabi or Saudi Arabia, but all of them share the characteristic that their development for many years ahead is unlikely to be constrained by shortage of capital or foreign exchange. Revenues accruing to them from the export of petroleum are likely to render worries about such matters superfluous.

For the purpose of analysing trade matters it seems sensible to group nations by their principal exports, as has been done for the OPEC members, but it must also be recognised that the structure of their imports is very important in influencing their welfare. Their degree of dependence on imports of oil, food and fertilisers for example has become a key issue.

Structure of Exports

While it is both true and important to recognise how crucially dependent upon exports of primary commodities the developing countries are, it is worth noting that in terms of non-oil primary commodities the rich countries still export more than do developing countries, as shown in Table 5.1

The figures in Table 5.1 are for total commodities. Now that oil prices have shot up to over five times 1970 prices, the absolute value of primary commodities exported by what are still called developing countries has greatly increased. If oil is excluded, though, the developed countries are the larger exporters of primary commodities, particularly food and fertilisers. This is why the popular assumption that a boom in commodity prices will benefit the developing world may be false. It depends on which commodities benefit most. Actually, in 1973 some thirty-one developing countries, with more than 50 per cent of the population of poor countries, suffered a deterioration in their balance of payments as a result of the commodity boom.[3] Note that this preceded the oil crisis which has also, as is widely known, hit most severely at the current-account balance of payments of a number of large and very poor developing countries.

The rise in commodity prices, excluding oil, is shown in Table 5.2. It also shows something of the variety of experience between groups of commodities. Even within the groups experience between commodities has been disparate. Among the food items for example tea and coffee have done much less well than cocoa.

wheat or sugar, as shown in Table 5.3. These contrasts in experience have to be borne in mind when making judgements about the commodity boom of 1973–74.

The sheer volume of primary commodities as a share of total exports in most developing countries inevitably connects their

TABLE 5.1
Value of Primary Commodities and Manufactures as a Percentage of Total Exports, 1960 and 1970

	Exports ($1,000m)		*Percentage of total exports*	
	1960	*1970*	*1960*	*1970*
Developing countries				
Primary commodities[a]	22.0	41.2	85.0	75.9
Manufactures[b]	3.8	12.7	14.6	23.4
Miscellaneous[c]	0.1	0.4	0.4	0.7
Total exports	25.9	54.3	100.0	100.0
Developed countries				
Primary commodities[a]	25.3	51.4	30.8	22.9
Manufactures[b]	54.0	169.1	65.9	75.4
Miscellaneous[c]	2.7	3.7	3.3	1.7
Total exports	82.0	224.2	100.0	100.0

[a] Standard International Trade Classification (SITC) 0–4.
[b] SITC 5–8.
[c] SITC 9.
Source: *Trends in Developing Countries* (Washington: International Bank for Reconstruction and Development, 1973).

TABLE 5.2
World Commodity Prices[a]

	January 1972	*3 January 1973*	*2 January 1974*	*8 January 1975*
Food	100.7	165.9	244.9	288.8
Industrial raw materials	94.1	123.8	203.3	145.0
fibres	116.1	190.0	275.0	195.7
metals	82.8	85.6	163.4	117.8
All items	101.3	147.2	226.4	224.9

[a] In United States dollars with 1970 = 100.
Source: *The Economist*, London, January 1974 and January 1975.

TABLE 5.3
Food and Beverage Prices[a]

	1971	*1972*	*1973*	*1974*[b]
Coffee	83	94	123	129
Cocoa	78	94	187	307
Tea	96	96	97	131
Sugar	105	112	127	301
Corn	102	96	168	226
Wheat	107	124	242	300
Soyabean oil	104	91	168	294

[a] Wholesale prices, 1970 = 100.
[b] Provisional, based on averages of first 7–8 months.
Source: *International Financial Statistics* (Washington: International Monetary Fund, November 1974).

prospects for earning foreign exchange with the fate of primary goods. But manufactured exports from developing countries have been growing fast. From 1965 to 1972 exports of manufactures from developing countries achieved an average rate of growth of 19.5 per cent per annum compared with 14.1 per cent for developed countries. There are a number of developing economies whose key to continued economic growth is the expansion of manufactured exports. It is not true that success in this field has been confined to a few special cases in South-East Asia. A number of large developing countries had already shown rapid growth in exports of manufactures by 1968 as is shown in Table 5.4

There is also a number of countries favoured by climate, history and geography that have gained considerable amounts of foreign exchange from tourism. Mexico, Turkey, North African countries and Kenya have been especially fortunate.

Structure of Imports

As Table 5.5 makes clear, capital equipment and manufactures form over 50 per cent of total imports for India, Pakistan and Indonesia. Nevertheless food (mainly grains), fuel and chemical fertilisers, even at the low prices prevailing in 1971, made up over 25 per cent for India and Pakistan. Indonesia is of course a net exporter of oil. The rise in the price of grains, shortages and high prices of fertilisers, insecticides and oil hit India, Pakistan and Bangladesh very hard in the early 1970s, with the cost of oil

TABLE 5.4
Population, Manufactured Exports and Growth of Manufactured Exports in Selected Developing Countries

	Population (m) 1969	*Manufactured exports ($m) 1968*	*Growth rates of manufactured exports 1958–68 (% per annum)*
India	526	936	6.6
Pakistan	127	364	30.0
Brazil	92	217	14.7
Mexico	49	244	8.4
Philippines	36	127	21.0
Korea (Rep.)	31	338	61.4
Argentina	24	298	8.0
Morocco	15	73	30.0
China (Taiwan)	14	606	28.0
Hong Kong	4	1602	14.5
Singapore	2	106	14.4

Source: Selected from a list of 45 developing countries which export manufactures, given by Hollis Chenery and Helen Hughes in "International Division of Labour: The Case History of Industry", paper (mimeo.) for conference of the Society for International Development, Washington, October 1971.

having risen to $1,350m in the Indian balance of payments in 1974. Many developing countries are in fact vulnerable to high prices for these primary commodities as an examination of the effects of the commodity boom will show.

Direction of Trade

By far the greatest part (over 70 per cent by value) of exports from developing countries goes to the developed market economies of the Organisation for Economic Cooperation and Development (OECD). Only about a fifth of their trade is with other developing countries. Although this and trade with socialist economies are growing fairly rapidly, they may remain small as a proportion of total exports, as shown in Table 5.6.

Clearly, despite the proliferation of customs unions and free-trade associations among groups of developing countries in the 1960s, trade among them remains quantitatively small relative to their trade with the industrial nations. Qualitatively this is likely

TABLE 5.5
Analysis of 1971 Imports

SITC		India ($m)	India (% of total imports)	Pakistan ($m)	Pakistan (% of total imports)	Indonesia ($m)	Indonesia (% of total imports)
0	Food and live animals	284	12	125	13	99	9
2	Crude materials except fuel	282	12	42	5	22	2
3	Mineral fuels	245	10	80	9	29	3
4	Animals, veg. oils and fats	51	2	46	5	–	
5	Chemicals	291	12	106	12	138	13
6	Basic manufactures	581	24	197	21	326	30
7	Machines, transport equipment	612	25	301	32	439	40
8	Misc. manufactures	43	2	24	3	44	4
9	Others	16	–	–	–	–	–
Total imports		2405	100	921	100	1097	100

Note: Columns do not add up precisely because of rounding.
Source: *Yearbook of International Trade Statistics 1972/73* (New York: United Nations, 1973).

TABLE 5.6
Direction of Commodity Exports from Developing Countries in 1970

	World total	*Developed countries*	*Developing countries*	*Socialist countries*
($m)	55	40.3	11.2	3.1
(%)	100	73	21	6

Source: *Handbook of International Trade and Development Statistics* (New York: UNCTAD, 1972).

to be even more true in terms of the difficulty in obtaining many capital goods and sophisticated manufactures from other developing countries at competitive prices. The situation is changing and will continue to do so, but it remains true that relatively small percentage changes in the value of trade with developed countries can have a big impact on the balance of payments, while it requires a much larger percentage change in the value of trade with other developing countries to significantly affect the foreign balance. The prospects for growth in exports are likely to remain heavily dependent on the ability to increase the value of exports to the developed market economies, either through increases in volume or prices of exports.

EVENTS OF THE EARLY 1970S

Having sketched in briefly in the forgoing some of the main characteristics of the trade of the developing countries, the impact of the main events from 1970 to 1974 can be considered against this background. These events can be listed as:

(a) world wide inflation, but concentrated in the industrially-developed nations of the OECD;

(b) international monetary crises and exchange-rate variations;

(c) the commodity boom and shortage of food and fertilisers; and

(d) the oil crises.

Although these events are connected, it is as well to examine them separately and then to note how they are linked. In that way a clearer picture of the timing and the direction of causation should emerge.

Throughout the previous decade the world as a whole grew at about 5 per cent per annum. From 1961 to 1972 the OECD nations as a group achieved a 5.1 per cent per annum growth of gross domestic product (GDP). These were unprecedented growth rates and they were not without their strains. Most countries were operating at or near the limits of their capacity and most were experiencing rather rapid rates of inflation, with consumer prices increasing by at least 4 to 5 per cent per annum. Then in 1972 and 1973 a simultaneous upward swing in production raised the growth of OECD countries to an unsustainable rate of over 6 per cent for 1973 as a whole, and reaching a rate of 8 per cent in the first quarter of 1973 and were running at a rate of 10 per cent by the end of 1973, despite the slackening of growth which came then. Thus inflation in Western countries was not caused by either the commodity boom or the rise in oil prices, but was itself in fact a cause of these events.

The surge of demand in 1972–73, pressing against inelastic

supplies of primary commodities, was probably the main cause of the dramatic increase in commodity prices, but other factors also played a role. The devaluation of the American dollar and the progressive abandonment of fixed exchange rates increased uncertainty about the value of currencies and, together with inflation, stimulated speculative holding of commodity stocks. The Japanese government encouraged firms to hold stocks of imports in an attempt to cut Japan's huge balance-of-payments surplus which was drawing critical attention from her OECD partners.

In addition to these factors there was another systematic cause of the rise in prices. Production of most agricultural products in 1972 was below trend for both general and specific reasons. Many developing countries for various reasons had not expanded output of agricultural products. This was partly a general neglect of agriculture in favour of import-substituting industries. High levels of protection for these industries, combined with controls on food prices and taxes on agricultural exports, lowered incentives to produce many primary products. In general, low domestic rather than international prices were responsible. International prices of some commodities such as "free market" sugar and rubber had lagged behind prices of manufactures in the 1960s, but on the whole the commodity terms of trade were fairly constant throughout the decade 1961–71 (see Table 5.7).

TABLE 5.7
Developing Countries' Terms of Trade
(1963 = 100)

1961	*1962*	*1963*	*1964*	*1965*	*1966*	*1967*	*1968*	*1969*	*1970*	*1971*
102	98	100	101	99	101	100	101	101	101	101

Source: United Nations.

Another influence on the situation was the growing shortage of fertilisers and pesticides. Underestimation of the expansion of world demand which accompanied the spread of new varieties of wheat, maize and rice, led to delays in building chemical plants. It takes about five years to build a fertiliser factory. Environmental lobbies in the United States and other countries opposed the location of plants within their countries and it became almost impossible to build new plants in America. As a result a ton of urea, which cost $62 in 1968, cost $225 in 1973.

1972 saw the arrival of some very specific factors. The Soviet wheat harvest was catastrophic, and the USSR entered the rather narrow market with huge demands. China unexpectedly purchased significant amounts of agricultural raw materials. The movement of the Humboldt current ruined the fish harvests of the Peruvian fishmeal industry, pushing demand for cattle fodder onto soya cake. Demand for cereals was growing rapidly for two main reasons: (a) the growth of population and income in developing countries, and (b) the rapid growth in demand for grain-fed meat in Japan and other rich countries.

The combination of these specific factors with the general situation of high demand and reduced supplies, pushed agricultural prices through the roof: sugar and rice tripled, wheat quadrupled.

The general increase in industrial demand explains the rise in prices of raw materials, both mineral and agricultural, but here too there were specific supply factors. The cotton crop in the United States was well down in 1972. Copper supplies were restricted by the political troubles of Chile and by the closure of the Rhodesian border with Zambia (Chile and Zambia are two of the world's major copper exporters).

The result of this complex of long-run, short-run and random factors was a rise in commodity prices of unprecedented magnitude, much greater than the Korean War boom of the early 1950s.

Increase in Oil Prices

Long before the War of *Yom Kippur* the prices of oil had started to rise, mainly because of the inability of the United States to continue to meet its energy needs from domestic sources. This plus growth of demand for energy in other countries produced, as discussed in Chapter 1, a surge in demand for oil so that the oil market moved from persistent surplus to shortage. The actions of OPEC had made little difference to oil prices up to the 1970s, but their bargaining power was strengthened by the growth in demand and also by increased solidarity in the face of the inflation in the prices of manufactures, which was reducing the real value of their oil revenues. Lybia, Iran and Kuwait became more aggressive in their handling of the oil companies and the growth of a dominant oil-supplier in the shape of Saudi Arabia set the scene for OPEC to become a successful cartel.

Posted prices rose from $1.80 per barrel, which had held from 1960 to 1970, to $3.01 on 1 October 1973. Estimated market prices over the same period rose from $1.30 to between $2.70 and $3.10 a barrel. This last was a situation where the market price was tending to exceed the posted price which determined tax revenues. Such a reversal of the normal relationship meant that the share of profits going to the companies was inadvertently rising relative to the governments' shares. This led to the governments' unilateral raising of the posted price to $5.12 on 16 October 1973 and agreeing to cut output by 5 per cent a month and to embargo exports to the United States and the Netherlands. On 1 January 1974 the posted price was raised to $11.65 and since then market prices have varied around that mark.

EFFECTS OF EVENTS ON DEVELOPING-COUNTRY TRADE

Clearly the key event of the first half of the 1970s was the dramatic increase in oil prices. The remainder of this chapter constitutes an attempt to discuss the effects of the events of the early 1970s on trade and trade prospects in developing countries.

Direct Effects of Higher Oil Prices

The direct effects of the oil price increases can be portrayed simply in terms of the impact on the current account of the balance of payments of net exporters and net importers of oil. Estimates vary on the distribution of the gains among the oil-exporting nations and the financial embarrassment likely to be produced for the rest of the world. The World Bank estimated the additional revenue accruing to the eleven major oil-exporting developing countries in 1974 over 1973 as $65,000m. Of this about half would have accrued to five very rich countries. Saudi Arabia, Kuwait, Abu Dhabi, Qatar and Libya, with a total population of about 11–12m. Clearly these countries cannot in the foreseeable future spend these enormous annual surpluses on imports. The repercussions of this fact on the world economy are considered below under the heading of indirect effects.

The extra cost of oil imports to developing countries in 1974 was $10,000m. The major importers of oil among developing countries have of course been hard hit. The impact happens to be worst among the most populous and poor: Bangladesh, India, Pakistan and Sri Lanka. Some estimates of the additional cost of oil imports are shown in Table 5.8 for a selected group of developing countries.

The largest consumer, Brazil, is likely to be less affected in the future for several reasons. First, Brazil is already producing one quarter of domestic oil needs. This is likely to expand and Brazil has considerable potential for hydroelectric power which is now very competitive with oil energy. Secondly, Brazil has large financial reserves and a fast growing manufacturing industry whose exports attained growth rates of 10–20 per cent in the first half

TABLE 5.8
Incremental Oil Import Expenditure of Developing Countries due to Effect of Price Increase since 1970[a]

	Actual		*Estimated*	*Projected 1980*		
Developing countries	*1972*	*1973*	*1974*	*low*	*medium*	*high*
Large developing countries						
India	73	203	1,090	2,895	3,600	4,290
Pakistan	18	42	210	385	475	565
Bangladesh	6	15	75	130	165	195
Selected other countries						
Uruguay	9	25	120	230	290	350
Turkey	32	77	425	960	1,175	1,400
Morocco	10	24	175	375	475	575
Ghana	4	11	55	120	150	180
Kenya	7	18	90	195	240	290
Sri Lanka	9	24	120	230	285	340
Philippines	45	118	580	1,110	1,390	1,665
Thailand	32	83	400	815	1,020	1,220
Exporters of manufactures						
Brazil	96	222	1,085	2,360	2,965	3,585
Korea	52	148	850	1,155	2,690	3,125
Argentina	11	15	60	335	425	515
14 selected countries sub-total	405	1,025	5,335	12,295	15,345	18,415
All developing countries	900	2,290	11,635	25,395	31,720	38,080

[a] The incremental expenditure is the difference in price between 1970 and the current year, multiplied by import volume in the current year.
Source: World Bank estimates, January 1974, reproduced in the *Institute of Development Studies Bulletin*, October 1974, p. 42.

of the 1970s. South Korea's buoyant exports of manufactures plus American assistance are likely to alleviate its problems. Argentina is relatively affluent and its grain and beef exports are likely to continue to do well.

Several Latin American countries are in fact receiving loans from Venezuela to aid adjustment to the new oil costs. This does not reduce the real burden. Resources have still to be devoted to the production of goods to pay off the increased oil costs and to service the loans, but the balance-of-payments problems are eased. Chile and Jamaica are in difficult positions. Depleted reserves and low copper output under the regime of Salvador Allende have weakened Chile's economy, but increased production and higher copper prices since then have helped. The prospects for copper are good provided the economies of the industrial nations return to higher growth rates once more. Jamaica's oil bill in 1974 was about a quarter of its export earnings. Tourism may be hit by recession in the West and increased costs of transport, but prospects for bauxite are good, and while the outlook for sugar is not hopeful sugar producers have made some windfall gains.

In short all developing countries which are net importers of oil have been affected severely by the rise in oil prices. The increased cost of oil in 1974 to well over $11,000m has wiped out the value of the flow of financial resources from OECD countries to the developing countries. Some like Brazil, Mexico and Malaysia–Brunei can find alternative sources of energy and will reduce their import bills in that way. Mexico may even become a net exporter once more. On the whole however developing countries are much harder hit by the oil crisis than are the rich nations. This is because they have fewer marginal uses of energy. America is able to save large amounts of energy merely by lowering thermostats and speed limits while for most developing countries any reduction in energy consumption is likely to hit transportation and production in industry and agriculture fairly quickly. The capacity to find substitutes for imported oil is a lot less for most developing countries because of shortages of capital and technology.

Special assistance with the short-term balance-of-payments difficulties of the developing countries has been forthcoming through several channels. Although OPEC nations have rejected a two-tier pricing system as a general solution to the difficulties facing developing countries, some contracts for oil supply have been made at special concessionary prices. Iran for example is selling crude oil to India and Pakistan on preferential terms. Loans, credit facilities and deferred payment terms have also been arranged between Arab oil-exporters and Arab nations which are dependent on oil imports. Venezuela has aided several Latin American oil-importers. Somewhat surprisingly the Witteveen special facilities in the International Monetary Fund (IMF), created with credits for over $3,500m from OPEC nations to help with the balance-of-payments difficulties of poor oil importers, had not been nearly fully utilised up to January 1975.

Even if loans are made available they do not reduce the real resource cost to oil-importers of the increase in oil prices. They will still have to export a much larger quantity of goods in order to pay for their oil imports and to service any associated loans. The principal advantage of loans is to enable the hard hit economies to continue to operate at a high level of capacity and to create through higher savings and investment the capacity to produce exports and import substitutes to gain the foreign exchange needed to pay for oil.

Indirect Effects of Higher Oil Prices

The indirect effects of higher oil prices are diverse and widespread. In this section they are grouped under three headings.

Repercussions of Retarded World Growth

The major risk to the export prospects of developing countries stems from the possibility of a major recession or even a depression in the industrial nations. This would severely cut their demand for raw materials of all kinds. Exporters of rubber, metals and fibres would all be hard hit. Food-exporters would suffer less, since the income elasticity of demand for basic foods is not so great, and demand would be supported by continued population growth.

The risk of a depression is real. The sudden increase in the cost of oil imports has the deflationary impact of a heavy increase in taxation without matching government expenditure. In addition fears of foreign economic and political influence stemming from dependence on oil imports are leading nations into attempts to reduce oil consumption and invest in high-cost alternative sources of energy. These policies are inimical to prosperity and high rates of growth. Attempts to lower the level of domestic demand, to reduce inflation and the non-oil deficit, have multiplier effects on other economies and can lead to widespread unemployment.

The rich nations are increasingly aware of these risks and have a recent history of successful collaboration in trade and finance, which makes one hopeful that a depression will be avoided. President Ford's turnaround, from emphasising policies to combat inflation in America to lowering taxes to stimulate demand, was considered a most hopeful sign. The general worries about inflation have however delayed counter-recessionary action to a perilously late date in most of the Western economies. Reflationary policies operate with considerable time-lags, and fears that they will generate further rises in wages and prices make governments nervous of taking forceful action to expand demand.

Figures both from OECD sources and from the National Institute for Social and Economic Research in Britain show the dramatic slowing of the growth of the major industrialised nations in 1974. Their forecast of recovery to positive though low growth in 1975 was considered to be if anything over-optimistic. Industrial production fell in almost every major industrial economy save France, and the growth of gross domestic product appears to be negative or relatively low for all (see Table 5.9). According to the same source unemployment rose throughout 1974 in all major OECD countries. According to press reports this trend to higher unemployment was expected to continue through 1975, and cer-

TABLE 5.9
Annual Rates of Growth (% per annum)

	1962–72	*1973*	*1974 (forecast)*	*1975 (forecast)*
United States of America	4.2	5.9	−2.0	0.7
Canada	5.5	6.8	4.0	3.3
Japan	10.4	11.0	−3.0	4.5
France	5.7	6.1	4.5	3.5
Germany (FR)	4.5	5.3	1.5	2.5
Italy	4.6	5.9	3.1	1.7
United Kingdom	2.7	5.2	−0.3	1.6
Others	5.0	5.4	3.4	3.2
Total OECD	5.1	6.3	−0.1	1.9

Source: *National Institute Economic Review*, November 1974, p. 25.

tainly in Britain it was widely expected to rise from its January 1975 level of 750,000 to over a million out of work before the end of that year.

For the longer run the outlook is less gloomy. Agreement in the Witteveen–Healey[4] proposals for a further and larger oil lending facility to recycle oil surplus revenues through the IMF to industrial nations in balance-of-payments difficulties due to oil imports, has significantly improved prospects for avoiding depression and renewing growth. The existence of a substantial (over $6,000m) special facility (though much smaller than the $25,000m proposed by Healey) does much to reduce the risks that sudden capital flights between nations will threaten monetary stability. If another $20–25,000m fund can be established along the lines suggested by Henry Kissinger,[5] the liquidity problem may be largely resolved.

In principle the extra costs of oil and the debt repayment likely to accumulate by 1980 are not too serious for the OECD nations, as has been cogently argued by Hollis Chenery of the World Bank in an article in *Foreign Affairs*.[6] At its peak the likely debt service due to oil loans should be less than 2 per cent of combined OECD gross national product (GNP). As he says, "It is difficult to argue on economic grounds that the world economy cannot sustain capital flows of the required magnitude or that the OECD countries need suffer heavily in the process".[7] The difficulties may in any case be reduced by some easing of the real cost of oil as the 1980s are approached. By the mid-1970s oil prices were above the costs of potential substitutes: nuclear energy, hydropower, shale oil and tar sands. They were also sufficiently high to generate a scramble to find oil which was turning up new oil fields every few weeks. This was seen as a potential threat to the solidarity of OPEC which could lead to some softening of prices. In the absence of some major political event, such a weakening of oil prices was not likely to be very significant until nearer 1980.

It was thought that if a depression were avoided and 1975–76 saw the resumption of growth by the OECD nations at an average rate of around 5 per cent per annum the outlook for most developing countries' exports would be vastly improved, but there would be a difficult period from early 1975 until 1976 for most raw material exporters.

Synthetic Costs

Petroleum provides the main feedstocks for the production of plastics and synthetic fibres. Since these compete with textiles, hides, timber and metals produced in developing countries, some benefits arise for the exporters of the natural products such as cotton, jute, hard and soft fibres, hides and non-ferrous metals. This is certain to be the case, but it may not be very important on its own. The proportion of final cost represented by the derivatives of petroleum in the finished products tends to be rather small and cannot result in a very large rise in their prices. Perhaps the effect of inflation on both labour and capital costs of the petro-chemical plants may be more significant in raising their product prices relative to imported natural raw materials. Certainly the recognition in OECD nations of the high social costs of large petro-chemical plants in crowded areas of their countries is militating against their expansion and raising their costs. It is increasingly difficult for them to find locations for their plants in the face of well-organised opposition by the environmental lobbies in OECD nations, and existing plants are having to install costly anti-pollution systems.

Another element in the picture is the increased cost of transportation due to higher oil prices. It is difficult to say how important this is, but most raw materials are high bulk in relation to value, so that transport costs from plantation or mine to final market may be a relatively high proportion of final delivered prices for some. This will militate against their ability to compete with synthetics normally produced nearer the industries of the OECD nations. On the other hand it may, along with the pollution factor, stimulate an increase in the trend to locate processing industries closer to mining operations. Processing at site is normally energy-saving and most developing countries are prepared to tolerate the adverse side effects of copper smelters and the like for the sake of employment and the increase in local value added which they bring.

Investment in plants to carry out the first stages of processing may represent an attractive possibility for combining OPEC capital with OECD technology and developing countries' labour: a rather appealing solution to some of the problems of all three groups.

Fertilisers, Pesticides and Energy for Agriculture
The high prices of fertilisers and pesticides in 1973–74 were more the result of the pressure of demand to expand agricultural output simultaneously in many countries, bearing on an inadequate capacity for production of these chemicals, than of the increasing cost of oil. Now however high energy costs and continued high demand will ensure that prices remain high. High demand in countries desperate to expand food output together with the normal time-lags in constructing fertiliser plants will also support high prices. The price of natural phosphates rose almost as much as oil prices in 1973 and 1974 and seems likely to hold up. Agriculture is of course increasingly dependent on energy inputs not only through chemical fertilisers and pesticides but in the means of applying them, in the supply of irrigation water through pumps and tubewells, and in tractor cultivation and transportation. The combination of high costs in all of these is exceedingly damaging to countries like India, which are struggling to expand agriculture by means of technologies, such as new seeds, which demand high energy inputs through water, fertiliser and plant protection.

OUTLOOK FOR OTHER COMMODITIES

Developing countries' main source of revenue is very often the export of primary commodities. But as with oil, many commodities assume even greater significance as imports. Food and fertilisers, both chemical and natural, loom large in several countries' import bills.

Food and Fertilisers

For the vast majority of the populations of developing countries the rise in the price of food and fertilisers may be much more important than the rise in oil prices. The oil price increase added about $11,000m to the import bill of developing countries compared to an increase of about $6,000m increase for food and fertilisers, but the latter affects the poorest people in the poorest countries most severely. India, Pakistan, Bangladesh and Egypt are all dependent on imports of food and fertilisers. Shortage of cultivable land combined with population growth rates of around 2.5 per cent per annum have made increased productivity of land essential. Chemical fertilisers have been a major source of this increased productivity. The problem has been exacerbated in Egypt by the effect of the Aswan Dam in reducing the spread of fertile silt over the traditional Nile lands. All those developing countries which run a significant food deficit have suffered. Nor is this situation likely to improve soon. As argued above, fertiliser costs are likely to stay high for several years and the same is likely for basic foodstuffs such as wheat, rice, maize and corn.

The huge stocks of grain which America once held are all gone and the continued weakness of Russian agriculture is a constant threat to the stability of the market because of Russia's sheer size. World exports of grains are very small in relation to world production so that quite a small percentage decline in food output in the Soviet Union can mean that the Russians buy up a very large percentage of the wheat available for export. World stocks will have to be rebuilt in order to be able to take care of such large fluctuations in the market so that even if the major grain-suppliers increase production significantly a substantial part of the increase ought to be used to increase world stocks so as to avoid a potential disaster. This should prevent any decline in grain prices. World Bank projections for 1975 show grain (especially wheat) prices remaining very high.[8] Argentina is the only developing country which stands to gain much from the shortage of wheat. Burma and Thailand should gain something from increased rice prices.

Argentina will also gain from high prices for beef, oilseeds and wool. Uruguay benefits from beef, and Peru from fishmeal. The main gains to developing countries from high food prices have probably been in sugar, cocoa and coffee. The Philippines, Brazil, the Dominican Republic and several Caribbean nations export sugar and they have benefited. But most sugar is sold under long-term agreements which have not followed the rise in free market prices. The free market is actually rather small. Moreover sugar prices are unlikely to be sustained at these levels for long. The cost of production of sugar is well below these levels and the high prices are the result of a lagged response of supply to low prices in the 1960s. If the problems associated with beet production in the European Community[9] were overcome, sugar exporters' hopes would have to be pinned on new and better international agreements, or changes in agriculture in the community and elsewhere.

The tropical beverages, coffee, cocoa and tea have not done as well as other food crops, nor are their prospects very appealing. Coffee and tea in particular face grave difficulties in expanding their markets since income elasticities of demand for them are low and their main markets are close to saturation. Eastern Europe, Russia and Japan are the main hopes for increases in coffee exports. For tea, only expanded consumption in developing countries holds out hope for increased sales. Cocoa's prospects are better. Both in OECD countries and in the Soviet Union consumption of cocoa has been growing quite fast.

Fats and oils have done moderately well, but most of the gains have gone to soya beans which are exported mainly by the United States. The sharp rise in fishmeal prices is partly due to the general factor of growing demand for protein feedstuffs for cattle and poultry which has helped all the fats and oils, but is mainly explained by the failure of the fish to appear in Peruvian waters. With the recovery of the Peruvian fish exports prices will moderate.

A general projection for all beverage and food exports is attempted in the World Bank study referred to earlier. In real terms the 1980 figure is substantially below the 1975 price level because of rising prices for manufactures to be expected over the period. According to the study the gainers among developing countries include mainly Argentina, Brazil, Uruguay, Ghana, Nigeria, Philippines and the Ivory Coast. The principal losers are India, Bangladesh, Pakistan, Sri Lanka and some of the poorer African countries. Several North African countries, especially Morroco and Tunisia, have gained and should continue to do well from the high prices of phoshates for fertilisers. A few developing countries with surpluses of natural gas may in future gain from exports of urea for fertiliser, but they will most likely be also exporters of oil.

Agricultural Raw Materials

Quantitatively the most important agricultural raw materials for developing countries are cotton and rubber. Egypt, Mexico, Sudan and Brazil being major exporters of cotton, it is envisaged that price rises would benefit all. The increased costs of synthetics should bring some gains to all of the fibres, and the use of cotton in blended fabrics with synthetics is growing rapidly and should sustain demand. The prospects for rubber are difficult to gauge.

The sharp decline in demand for cars and in the mileage covered by automobiles as a result of high costs of petroleum have cut demand in the major market for rubber. On the other hand, high oil and tallow prices raise the costs of synthetic rubber production which competes closely with the natural product. The World Bank projection puts the 1980 price fairly high. Wool should do well and this will benefit Argentina.

The other principal fibres, sisal and jute do gain from the effect of increased petroleum costs upon polypropylene production, synthetic rubber and polythene. All of these synthetics compete with them in certain uses. However general trends in packaging and in agriculture are probably against them. They will do well if they attain the prices projected in the World Bank study.

Timber is quite another matter. The world's demand for wood pulp and for quality woods in furniture is growing rapidly, and prices should go on rising as many of the accessible supplies of timber are being used up faster than they can be replaced. The major suppliers are however developed countries such as Canada, Sweden and Finland. Many developing countries do export valuable hardwoods, and the prospects for developing fast-growing tropical softwoods seems good.

The world tobacco market is expanding slowly, but demand is switching from the American tobaccos with their high nicotine content to types exported from developing countries. Looking at this series of agricultural raw materials it would seem that the gainers will again be Argentina from wool and Brazil from cotton with Egypt, Mexico and the Sudan also gaining from cotton prices. Timber and pulp-exporters will do well too, but these are not major exports from developing countries, and gains in tobacco are likely to go mainly to Greece and Turkey, rather than the poorer countries.

Bangladesh is doubly unfortunate in that its major export, jute and jute goods, gain relatively little from the high costs of petroleum and fertiliser which have severely hit her import bill.

Minerals and Metals

Copper is the most important of these, and it crucially affects the economies of Chile, Zambia, Zaïre and Peru. Prices rose sharply from a very low level in 1972 to almost double that in 1973 and continued to rise until May 1974, but then they returned to below 1973 levels. The increase was partly due to demand factors, but disruption of Chilean supplies and temporary difficulties in transport for Zambia were major causes. The price level is unlikely to keep up with the rising prices of manufactures. Zinc, bauxite and tin should do much better. These should benefit Mexico (zinc); Jamaica, Surinam, Guyana and Guinea (bauxite); and Malaysia, Bolivia, Thailand, Indonesia, Nigeria and Zaïre (tin). Mexico and Peru export lead, but this is not likely to bring great gains as it is unlikely to keep up with rising prices of manufactured imports. The same is true, but rather worse, for manganese, which is exported by Gabon, Brazil and India.

From this brief survey of the outlook for commodities it is apparent that while they might do much better than was thought in the 1960s (with the exception of petroleum), the gains of 1973 and 1974 are not likely to continue. Moreover the gains of some developing countries have been the losses of others. Worse still the commodity prices which have risen most and have good prospects are ones in which the OECD countries are the main exporters.

Prospects for Other Commodity Cartels

There are many commodity exports from developing countries whose prices are likely to fall behind the prices of their imports. Given the example of the success of OPEC in unilaterally raising oil prices, many producers must be examining the possibility of emulation.

The possibility of achieving a sustained price increase for many primary commodity exports from developing countries is slender and run risks of retaliation from importing nations. After a careful study of various possibilities, John Pincus concluded that only coffee, cocoa, tea, sugar and bananas were reasonable candidates for international commodity agreements designed to restrict supplies and raise prices.[10] Since his study, the International Coffee Agreement has come and gone. During its "heyday" it may have transferred half a million dollars a year to producer nations – hardly a success story. After many years of negotiation an International Cocoa Agreement was set up, but by 1975 had done nothing to affect markets. The International Sugar Agreement expired in 1973, and the new agreement negotiated in Geneva at the end of 1973 contained no economic provisions. Turning to minerals, the only agreement to have been sustained over a lengthy period is the tin agreement, which dates back to the 1930s. No producer organisation other than OPEC had by 1975 succeeded in achieving significant price increases for exports of developing nations.

Oil has certain key attributes which make it particularly suitable for a cartel:

(a) It is an essential requirement of major industries, transport and heating in OECD nations. Because of years of relatively cheap energy from this source, developed countries have become extremely dependent on it. In the short run there are few possibilities of finding substitutes for it. This gives it the important characteristic of being highly inelastic in demand.

(b) It can be simply left in the ground without deterioration and without causing serious unemployment to workers in the producing countries. There is no need for costly stockpiling.

(c) Several major exporting countries have enormous reserves of foreign exchange and huge surplus revenues over their requirements for imports. They can easily afford to restrict output even if this reduces their income, and they can face economic retaliation with aplomb.

(d) A significant group among them, the Organisation of Arab Petroleum Exporting Countries (OAPEC) nations, share a common political objective and have cultural and religious affinities which enhance their solidarity within the organisation.

Probably no other commodity of importance in world trade displays all of these advantages. Some commonly mentioned possibilities are copper, bauxite and tin. Compared with oil's share in world exports in 1970 of 5.04 per cent they are rather small: copper 1.29 per cent, aluminium (bauxite is the raw material for most aluminium) 0.67 per cent, tin 0.21 per cent.

To some extent they are substitutes for one another, but they also face competition from a variety of sources. It ought to be remembered that the United States is the world's largest producer of copper and lead and is a major producer of zinc. Recycling is the source of a great deal of non-ferrous metals in OECD nations, and this recovery of scrap is very sensitive to price. Plastics and glass compete with non-ferrous metals in many uses. The technological capacity of the industrial nations to respond to high prices for raw materials is very great. They can open up new mines and reopen old ones in their own countries. Previously uneconomic sources – such as low-grade ores, clay for alumina, seabed nodules for copper and manganese become viable. They can improve the efficiency with which existing materials are used. Telephone cables today use about 25 per cent of the copper thought necessary decades ago. For copper, actual imports are only a small proportion of OECD countries' consumption so that it would only be a matter of cutting marginal uses to significantly reduce demand. Bauxite, copper and tin compete in a number of major uses so that cooperation between producers would increase prospects of success The cooperation however would involve rather a large number of countries with few common interests. The main copper-producing countries are Chile, Peru, Zambia, Zaïre, Canada, Australia, America and the Soviet Union; for bauxite, Australia, Jamaica, Surinam, Guyana and Guinea; for tin, Malaysia, Bolivia, Thailand, Indonesia, Nigeria and Zaïre. Besides these there are many minor exporters and very large numbers of countries with the potential in old mines and lower-grade ores. Altogether, it seems rather doubtful that successful producer cartels could be formed for these minerals. Yet on technical and economic grounds they are among the best prospects.

The general problem facing most cartels for primary commodities is the presence of good existing or potential substitutes in the medium term in the developed countries. Exceptions to that are the products picked out by John Pincus: tropical beverages, bananas and sugar, but previous attempts to control them have met with little success and, even on optimistic assumptions, they would be unlikely to transfer large financial resources to developing countries. There would be a risk that the industrial nations would cut their official aid by as much in retaliation.[11]

TRADE IN MANUFACTURES

This is the most dynamic sector of exports and, as shown in Table 5.4 above, affects over forty-five developing countries, many of them large nations. Of course it is still true that a few countries such as Hong Kong, South Korea, Taiwan, Israel and Mediterranean countries predominate, but others such as Brazil, Columbia and Pakistan are catching up; and India has the capacity to increase its manufactured exports enormously. The change in attitudes from "inward-looking", import-substituting development to an "outward-looking" export oriented path has stimulated highly successful growth in manufactured exports from several countries.[12]

Obstacles

Obstacles in the form of high effective tariff rates, quotas and more subtle non-tariff barriers are endemic in the markets for developing countries' exports of manufactures. Discriminatory trading systems such as the European Economic Community (EEC), the European Free Trade Association (EFTA), and arrangements between the United States and Canada also militate against them.

Over the last twenty years tariffs on manufactures have been cut drastically. The average tariff imposed by the United States is now less than 10 per cent. Unfortunately this average conceals wide variations, and in fact tariffs in America as in other developed nations bear much more heavily on the types of goods which developing countries normally can manufacture and export, such as textiles, footwear, leather goods, stone, ceramic and glass products. Moreover the tariff borne by the processing stage is often very high and discriminates against developing countries because the value added in early processing is often small (Table 5.10).

TABLE 5.10

Averages of Nominal and Effective Tariffs on Industrial Countries' Imports of Manufactures

	Nominal	*Effective tariff*
Total imports	6.5	11.1
Imports from LDCs	11.8	22.6

Source: Ian Little, Tibor Scitovsky and Maurice Scott, *Industry and Trade in Some Developing Countries: A Comparative Study* (London: Oxford University Press, for OECD, 1970) p. 273.

In the face of these tariffs the achievements of some developing countries in the late 1960s and early 1970s have been remarkable. Several have been assisted by preferential arrangements, by Commonwealth, EEC and special arrangements with America.

Also, since 1970 many industrial countries have adopted some form of a Generalised System of Preferences (GSP), first debated at the United Nations Conference on Trade and Development (UNCTAD) in 1964. In most cases however they have excluded goods of special importance to developing countries such as textiles, leather, rubber and wood products. In many countries low ceilings have been fixed on preferential imports. After the ceiling has been reached the most-favoured-nation tariff applies once more.

An examination of the effects of the GSP schemes implemented up to 1972 concluded that they brought negligible benefits. "The revenue transfers flowing from the programme are estimated to be less than $100m."[13] Actually, since the study by Tracey Murray, the system has liberalised somewhat and the potential benefits to developing countries which can diversify their range of exports are high.

Quotas and voluntary export restraints are widespread and iniquitous. "There is almost no export, apart from a few non-ferrous minerals, of any less developed country which is not subject to quantitative restrictions in some developed countries."[14] There is some evidence which suggests that non-tariff barriers also discriminate particularly against the exports of developing countries.[15]

Another major obstacle to exports of manufactures is current and past commercial policies in developing countries themselves. The atempts to industrialise through import substitution behind high levels of protection, have in many cases created inefficient industries which are not only uncompetitive themselves but raise the costs of other industries. Over-valued exchange rates in many developing countries lower profits in exporting and so reduce incentives. The damaging effects of these policies have been thoroughly documented in several recent studies.[15] Changes in attitudes to exporting and realisation of the distorting effects of high protection have led to recent improvements in export performance.

TRENDS AND CONCLUSIONS

A major and hopeful advance in exports of manufactures is documented in a paper by Gerry Helleiner in the *Economic Journal*.[17] This is the result in particular of American and Japanese firms hiving off labour-intensive parts of their production processes and locating them in developing countries. A similar trend is appearing for industries which are being forced out of developed countries on environmental grounds. A third trend is the organisation by large retail houses, such as Sears Roebuck and K. Mart from America, of production of goods to their specifications in developing countries. All of these three approaches solve one of the most difficult problems facing exporters from developing countries. That is, the problem of meeting quality standards and overcoming the difficulties of breaking into new markets in the rich countries.

Provided the rich nations recover their nerve and restore OECD growth rates to an average 5 per cent, there are considerable gains likely for exporters of manufactures from developing countries. The negotiations in the Tokyo Round of the General Agreement on Tariffs and Trade (GATT), held in Geneva, hold forth some prospects of improvements in the GSP. Better control over safeguard procedures was also seen as a likely outcome of these negotiations. Both would reduce uncertainty in export markets and enhance prospects for developing countries' exports.

Other trends favouring developing countries' own manufacturing industries are a likely reduction in the numbers of immigrant 'guest workers' in industrialised countries. Social problems arising from their presence combined with Europeans' dislike of assembly line jobs is pushing firms to locate such activities in countries where wages are lower and labour less fussy.

All of these developments suggest that those countries which can adapt their economic structure to produce manufactured exports have very good trade prospects through the 1970s.

Oil prices are likely to remain for the rest of the 1970s at roughly the level reached in January 1974.[18] This brings huge benefits to a number of countries presently classified as developing. Several of them are in any case large and poor in terms of *per capita* income, Nigeria, Indonesia, Iran, Iraq and Algeria. They had about 230m people and an average *per capita* income in 1971 of about $200. At the same time high oil prices cause great difficulties to a much larger group of people in India, Bangladesh, Sri Lanka, Pakistan and several African countries.

The indirect effects of high oil prices are exceedingly hard to evaluate. The most important issue is the effect on world economic growth. This need not be serious. In principle the real resource burden on the OECD nations of paying for more expensive oil and the oil debt burden is not excessive. The dangers come from the shock effect of the suddeness of the increase coming on top of an already unstable world economy. Provided the rich nations and the oil-producers collaborate in finding sensible solutions to the liquidity problems of oil-importers, provided there are no panic movements of capital, alternatively central banks combine to offset them, world trade should recover. But trends made 1974 a year of recession with prospects of even higher levels of unemployment in 1975, certainly in Britain and America. It is hoped that the latter half of the decade will see the restoration of normal growth.

Given these conditions commodity exports from developing countries should do fairly well. Grain exports, beef, sugar, fish-meal and oilseeds should maintain high price levels compared with 1970. This will benefit Argentina, Uruguay, Peru and a number of countries in South-East Asia and Africa which export rice and vegetable oils. Once again however the gains to these nations are balanced by losses from high food prices for countries such as India, Bangladesh, Sri Lanka and Pakistan which are net importers of food. This is exacerbated by the high prices of fertilisers which are also likely to continue to 1980. These will benefit only a few developing countries such as Morocco and Tunisia which export phosphates.

Raw materials in general will gain somewhat from the effect of high oil prices on synthetics and plastics but probably not a great deal, because transport costs for bulky items are raised.

Cartels in some primary exports of developing countries may arise, but they are unlikely to be able to raise prices significantly in a sustained manner.

Manufactured exports should do well because of a number of trends summarised above.

The upshot of these rather rough judgements is that no generalisation about trade in the 1970s for developing countries is possible. The prospects are very different for different groups of countries. On balance it is the poorest countries with the largest populations which seem likely to do worst.

NOTES AND REFERENCES

1. See p. 176 below.
2. The author is grateful to Dr V. N. Balasubramanyam and other members of the economics staff seminar at the University of Lancaster for comments on an earlier version of this paper.
3. *The Impact of the Recent and Prospective Price Changes on the Trade of Developing Countries,* UNCTAD/OSG/52 (Geneva: United Nations, 1974).
4. Proposals for a special facility in the IMF were first advanced by Johannes Witteveen, as Managing Director of the IMF, at the meeting of the Committee of Twenty in Rome in January 1974. Denis Healey, as Britain's Chancellor of the Exchequer, elaborated on the proposals at the IMF annual meeting in September of that year.
5. The original proposals of Henry Kissinger, as the American Secretary of State, were made in November 1974. These led to the formation, in March 1975, of a "safety net" fund by the International Energy Agency.
6. Hollis Chenery, "Restructuring the World Economy", *Foreign Affairs,* New York, December 1974.
7. *Ibid.*
8. *Additional External Capital Requirements of Developing Countries* (Washington: International Bank for Reconstruction and Development, March 1974) Table 4 and Appendix Table IV.
9. The problems of beet production in the European Community, in the context of world markets, are discussed in Ian Smith, *The European Community and the World Sugar Crisis,* Staff Paper no. 7 (London: Trade Policy Research Centre, 1974).
10. John Pincus, *Foreign Aid and International Cost Sharing* (Baltimore: Johns Hopkins Press, 1965).
11. For a more detailed discussion of international commodity agreements, see Alasdair MacBean, *Export Stability and Economic Development* (London: Allen & Unwin, 1966) ch 12.
12. G. K. Helleiner, "Manufactured Exports from Less Developed Countries and Multinational Firms", *Economic Journal*, Cambridge, March 1973; Jagdish Bhagwati and A. Kreuger, "Exchange Control, Liberalisation and Economic Development", *American Economic Review,* Papers and Proceedings, New York, May 1973; and Chenery and Helen Hughes, "International Division of Labour: the Case History of Industry", paper (mimeo.) for a conference of the Society for International Development, Washington, October 1971.
13. Tracey Murray, "How Helpful is the Generalised System of Preferences", *Economic Journal*, June 1973.
14. David Wall, "Opportunities for Developing Countries", in Harry G. Johnson (ed.), *Trade Strategy for Rich and Poor Countries* (London: Allen & Unwin, for the Trade Policy Research Centre, 1971) p. 36.
15. Ingo Walter. "Non-tariff Barriers and the Export Performance of Developing Countries", *American Economic Review*, Papers and Proceedings, May 1971.
16. Ian Little, Tibor Scitovsky and Maurice Scott, *Industry and Trade in Some*

Developing Countries: a Comparative Study (London: Oxford University Press, for the OECD Development Centre, 1970) p. 273; and Bhagwati and Kreuger, *loc. cit.*, besides the individual country studies for the OECD Development Centre edited by Little, Scitovsky and Scott.

17. Helleiner, *loc. cit.*
18. While there have been signs of weakening in OPEC prices this may only be temporary as a result of recessions in OECD countries.

CHAPTER 6

Impact of the Oil Crisis on the Energy Situation in Western Europe

GEORGE F. RAY

The oil crisis that broke at the end of 1973 hit the countries of Western Europe especially hard.[1] Its initial shock, and the disruption it caused, underscored the role of imported oil in the economic life of those countries.[2] Yet this stark reality contrasted greatly with the position of only a decade or so before when indigenous sources provided the bulk of Western Europe's energy requirements. Any assessment of the West European situation must therefore begin with the question: How did Western Europe arrive at this state of affairs? And how, and indeed why, has it become so heavily dependent on oil?

Although demand had been rising substantially, the 1950s and the 1960s were characterised by an over-supply of crude oil. The discovery of enormous reserves in the Middle East, followed by smaller but still significant finds elsewhere, resulted in very plentiful supplies. These the oil-producing countries could only place in Western Europe. This was the only market where large volumes were required. Japan was still a small consumer, even though she was growing rapidly, and the United States was practically self-sufficient.

The major international companies, the so-called Seven Sisters, were joined in those years by a number of newcomers: smaller companies which have come to be called "independents", mainly American companies searching for low-cost sources in order to distribute the oil in the huge and apparently insatiable American market. The success of the independents in finding oil in the Middle East and Africa was followed by the introduction of this oil into the United States. The established interests in the United States however reacted and in 1959 imports were effectively limited.[3]

Newly-found oil therefore had to be sold elsewhere and Western

Europe was the obvious target. Fierce competition in Western Europe followed. Demand for lighter products, chiefly motor spirit, rose very fast. But there are other products that result from the refinery process and markets had to be found for them. Accordingly, various types of fuel oil were offered at very low prices, successfully competing with coal.

It is worth underlining this point with some salient facts. In the Federal Republic of Germany, the comparable price of household coal in Hamburg, adjusted for heat content, was 43 marks per ton higher than the price of light fuel oil in January 1960; this difference increased to 103 marks by January 1970 and to 162 marks by July 1972. Prices in other West German cities indicate similar increases. In Munich for example the differences were 31, 116 and 190 marks respectively. In the United Kingdom the price of industrial coal rose by more than 60 per cent from 1955 to 1970, whereas in the same period the price of medium fuel oil for industry only rose by just over 3 per cent. In Sweden where coal has never been of any great importance and oil competed with hydroelectricity, the price of fuel oil did not change at all between 1950 and 1970, and indeed some grades became slightly cheaper.

The price advantage for oil products was greatly supported by the convenience of liquid over solid fuel. The result was that oil quickly became the most important primary source of energy in Western Europe.

ENERGY IN THE EARLY 1970S

Western Europe's energy situation can best be analysed on the basis of the consolidated and comparable statistics published by the United Nations.[4] It is convenient to begin with an analysis of the 1970 situation. The statistics clearly demonstrate the trends that were developing in Western Europe over the previous decade. These trends continued unabated between 1970 and 1973.

Consumption and Production Trends

In 1970 Western Europe (including Yugoslavia)[5] consumed the equivalent of 1,351m tons of coal. As shown in Table 6.1, about one-third of this was solid fuel; 56 per cent, liquid fuel; 8 per cent, natural gas; and the rest, just under 4 per cent, hydro and nuclear electricity. Ten years earlier, in 1960, consumption was only 836m tons coal equivalent, consisting of 65 per cent solid fuel, 30 per cent liquid fuel, 2 per cent natural gas and 3.5 per cent hydro and nuclear electricity. The trend was quite plain; the share of coal was halved in the decade ending 1970, while that of oil almost doubled and that of natural gas, although still small, quadrupled.

Oil and gas contributed in 1970, in relative terms, roughly the same proportion to Western Europe's total energy needs as coal did ten years before. In this period demand for energy rose at an average annual rate of 4.9 per cent,[6] but within this coal usage fell by 2.0 per cent, consumption of oil rose by 11.7 per cent and that of natural gas by just over 20 per cent a year.

Western Europe already had in 1960 an energy deficit, amounting to 273m tons of coal equivalent, of which 230m tons was accounted for by oil and 43m tons by coal. The deficit almost trebled by 1970, to 794m tons of which 728m tons of coal equivalent were due to oil, but the coal deficit also rose to 62m tons. Energy production remained more or less static between 1960 and 1970, but its pattern changed. Coal production fell by some 120m tons and this gap was filled by the additional production of natural gas (90m tons coal equivalent, mainly in the Netherlands and from the British part of the North Sea) and by the increased production of hydro and nuclear electricity (20m tons).

Compared with North American levels, consumption of energy in Western Europe remains modest. In 1970 it was 35 per cent of the level in the United States; and in 1960 it was 33 per cent. This conceals large variations in energy consumption among the countries of Western Europe, ranging from 6,304 kg. per head in Sweden to 687 kg. per head in Portugal, these being 1970 figures.

The shift away from coal meant that Western Europe, which in 1960 still covered two-thirds of its energy needs from "domestic" production, produced in 1970 not more than about 40 per cent of its needs. Again, in these terms, the situation varies greatly by country. Denmark has no energy resources at all. But the Netherlands produces some 70 per cent of her consumption. The position of the larger countries is almost equally uneven. In 1970 production, expressed as a percentage of consumption, ranged from 18 per cent in Italy to 31 per cent in France and 55 per cent in both West Germany and Britain. (Between 1970 and 1973 oil was the most rapidly rising element in total energy, which means

Basic Energy Statistics for Western Europe

	Production	Consumption		Proportion of total energy requirements (%)					
				Solid fuel	*Liquid fuel*		*Natural gas*	*Primary electricity*	*Per capita consumption*
	1970 CE[a]	*1960 CE*[a]	*1970 CE*[a]	*1970*	*1960*	*1970*	*1970*	*1970*[d]	1970[c]
Austria	10.7	15.3	25.3	28	27	49	15	8	3.4
Belgium[b]	11.6	37.6	59.6	39	24	50	10	1	6.0
Denmark	–	12.9	28.8	12	56	89	–	–	5.9
Finland	1.2	7.3	19.6	20	48	74	–	6	4.2
France	59.3	110.8	193.0	29	30	60	7	4	3.8
Germany (FR)	174.0	203.0	317.0	41	20	50	7	2	5.2
Greece	2.9	4.4	11.2	27	64	70	–	3	1.3
Ireland	2.5	5.2	8.8	40	29	59	–	1	3.0
Italy	26.4	56.0	144.1	9	54	75	12	4	2.7
Netherlands	49.0	32.7	66.2	11	50	48	41	–	5.1
Norway	7.6	9.8	18.7	8	48	54	–	38	4.8
Portugal	1.0	3.3	6.6	20	61	70	–	10	0.7
Spain	16.0	25.1	49.9	32	32	61	–	7	1.5
Sweden	5.0	26.0	51.0	6	73	82	–	12	6.3
Switzerland	4.0	10.4	21.4	3	54	81	–	16	3.4
United Kingdom	164.0	258.0	299.0	51	24	42	5	2	5.4
Yugoslavia	21.4	16.1	29.3	56	11	23	4	7	1.4
Western Europe	557.0	836	1351	32	30	56	8	4	3.8

[a] All types of energy, in million metric tons coal equivalent.
[b] Including Luxemburg.
[c] In metric tons coal equivalent; comparable consumption in North America was 10.9 tons *per capita* in 1960.
[d] Hydro, nuclear and imported electricity.
Source: *World Energy Supplies* (New York: United Nations, 1973).

the percentages given here would have been considerably lower in 1973.)

European coal production, once the cornerstone of the Industrial Revolution and later development, has contracted significantly. By 1972 only Britain and West Germany mined for hard coal on a significant scale, both producing over 100m tons a year. The once considerable French and Belgian mines had by then reduced their output to 30m and 10m tons respectively, with Dutch output even less. Brown coal (lignite) mining remained an important source in West Germany, Yugoslavia and Greece and on a much smaller scale in some other countries (see Table 6.2).

Oil had become by about 1970 the most important source of energy in every West European country with the exception of Yugoslavia. In many, oil was meeting more than 70 per cent of all primary energy needs; such was the case in Denmark, Sweden, Switzerland, Italy, Finland, Portugal and Greece. Even Austria and Yugoslavia, which in 1960 could still satisfy a large part of their demand from domestic oil production (78 and 67 per cent respectively), had to import 60–70 per cent of their required quantities. Western Europe was poorly endowed with crude oil supplies. Total production amounted to no more than 20–25m tons and met just a small fraction of requirements (see Table 6.3).

Thus imports have had to meet increasingly large shares of growing demand. As Table 6.4 shows, these imported supplies have come chiefly from the Middle East and North Africa; in 1970 51 per cent came from the former, 32 per cent from the latter. Shipments from the Soviet Union accounted for no more than 4 per cent of West European supplies. The USSR was an important source for Finland and Yugoslavia and to a minor extent for Italy, Sweden and Greece. In a number of cases the West European importng countries, without exception very dependent on the Middle East and with at least half of them dependent, too, on North Africa, shipped large parts of their supplies from only one or two producing countries. France was mainly reliant on Algeria; West Germany, Italy and Switzerland on Libya; the United Kingdom on Kuwait; and Sweden on Nigeria. Iran, Saudi Arabia and Iraq were important suppliers in other cases.

Primary electricity generation, that is hydro and nuclear power, was another source of energy, although in most cases not a very important one. With respect to hydroelectricity geological and

TABLE 6.2
Coal Production in Western Europe, 1972

Hard coal[a]		*Brown coal*[a]	
United Kingdom[b]	119.5	Germany (FR)	110.4
Germany (FR)	102.7	Yugoslavia	30.2
France	29.8	Greece	11.6
Spain	11.0	Turkey	4.7
Belgium	10.5	Austria	3.8
Turkey	4.6	Spain	3.1
Netherlands	2.9	France	3.0
Other countries	1.7	Other countries	—
Western Europe	282.7	Western Europe	166.8

[a] Million metric tons.
[b] Strike reduced output by about 15m metric tons.
Source: *Quarterly Bulletin of Coal Statistics for Europe* (New York: United Nations, 1973).

TABLE 6.3
Crude Oil and Natural Gas Production in Western Europe

Crude oil, million metric tons, 1970		*Natural gas, thousand T cal*[a]*, 1972*	
Germany (FR)	19.6	Netherlands	491.5
Yugoslavia	2.9	United Kingdom	250.9
Austria	2.8	Germany (FR)	153.1
France	2.6	Italy	129.8
Netherlands	1.9	France	69.9
Italy	1.5	Austria	19.0
Others	0.4	Yugoslavia	13.7
		Others	0.6
Western Europe	19.6	Western Europe	1,128.5

[a] 1 Teracalorie = 1,000m Kilocalories = 39,680 therms.
Sources: *World Energy Supplies* (New York: United Nations, 1973). *Annual Bulletin of Gas Statistics for Europe* (New York: United Nations, 1973).

TABLE 6.4
Crude Oil Imports of Western Europe, 1970

	Million metric tons	Dependence on[a]			Dependence on individual producer	
		Middle East	*North Africa*[b]	*USSR*	*25 per cent or over*	*15 to 25 per cent*
Belgium[c]	30	59	26	2	–	Libya 22; Kuwait 17; S. Arabia 16
Denmark	10	60	14	–	Kuwait 28	Nigeria 19; S. Arabia 17
Finland	10	29	–	69	USSR 69; Iran 29	–
France	103	45	43	1	Algeria 26	Libya 17
Germany (FR)	99	33	49	3	Libya 41	–
Greece	4	87	4	13	Iraq 53	Syria 24
Ireland	3	86	–	–	Kuwait 30; S. Arabia 27	Iran 23
Italy	114	55	32	9	Libya 30	Iraq 20
Netherlands	63	63	21	–	–	Libya 21; Kuwait 18; Iran 15; S. Arabia 15; Oman 22; Venezuela 18
Norway	7	55	9		–	Nigeria 17; S. Arabia 17
Portugal	4	97	–	–	Iraq 54	UAE[d] 22; S. Arabia 16
Spain	32	52	28	–	S. Arabia 29; Libya 25	–
Sweden	12	50	30	6	Nigeria 27	Oman 17
Switzerland	5	22	74	–	Libya 59	–
United Kingdom	101	61	25	–	Kuwait 26	Libya 24; S. Arabia 15
Yugoslavia	5	67	–	33	Iraq 44; USSR 33	Iran 22
Western Europe[e]	603	51	32	4	Libya 25	–

a Percentage of total.
[b] Algeria and Libya only.
[c] Including Luxemburg.
[d] UAE = United Arab Emirates (Abu Dhabi, Dubai, etc.).
[e] Shares of the main producers: Algeria 7; Libya 25; Nigeria 6; Venezuela 4; Iran 8; Iraq 9; Kuwait 13; Oman 2; Qatar 2; Saudi Arabia 14; UAE 3; USSR 4.

climatic conditions favour just a few countries, while the others cannot use hydropower for the generation of electricity on any large scale (see Table 6.5). Norway, Switzerland, Sweden, Portugal, Yugoslavia, Austria and Spain were among those fortunate enough to be able to cover more than half of their electricity requirements by hydropower. In the other countries hydropower adds some useful quantities to the electricity generated from other energy sources, or is of minor national importance only. Its local significance however cannot be belittled. A case in point is the United Kingdom, where hydroelectricity constitutes less than 2 per cent of all electricity generated, but is concentrated in the North of Scotland, which relies on it almost entirely. It should furthermore be remembered that even where hydroelectricity is the almost exclusive source of generation, such as in Norway, its share in country's total energy supply was, in 1970, not more than about a third, leaving Norway dependent on other primary fuels, chiefly on oil, for two-thirds of her requirements.

The contribution of nuclear power towards meeting Western Europe's energy needs was still minimal in 1970. Even in the United Kingdom, with almost one-half of total Western European nuclear power production, only about one-eighth of all electricity was generated in nuclear stations. Nowhere else in Western Europe did nuclear power contribute more than 8 per cent to total electricity generation, and nuclear output in both France and West Germany was approximately one-third of that in the United Kingdom.

In 1970 Western Europe had to import almost 60 per cent of its energy requirements: almost all – 97 per cent – of the crude oil consumed, some 14 per cent of the coal, but only 3 per cent of the natural gas consumption. Almost everywhere close to 100 per cent of oil consumption was met from imported oil. The only countries where production of other energy fuels exceeded consumption were the Netherlands for natural gas, and West Germany for coal, with Italy being just about self-sufficient in natural gas. The countries with good hydroelectricity resources – Austria, Switzerland and Norway, as well as Spain – produced somewhat more electricity than they consumed and supplied their surplus production to their neighbours.

The supplementing of inadequate indigenous energy resources resulted in a varying, but important and growing, part of the foreign trade of most countries. In addition to the very large

TABLE 6.5
Hydro and Nuclear Electricity in Western Europe, 1972

Hydro-generation			*Nuclear generation*		
	GWh[b]	*per cent*[a]		*GWh*[b]	*per cent*[a]
Norway	67.6	99	United Kingdom	25.6	12
Sweden	53.8	75	France[d]	9.3	6
France[d]	49.4	32	Germany (FR)	9.1	3
Italy[d]	40.0	32	Italy	3.4	3
Spain[d]	32.7	52	Switzerland[c]	2.5	8
Switzerland[c]	29.3	88	Spain	2.5	4
Yugoslavia	18.0	54	Sweden	1.5	2
Austria	17.2	59	Netherlands	0.3	1
Germany (FR)	13.7	5			
Finland	10.2	44			
Portugal	7.1	81			
United Kingdom	4.3	2			
Greece	2.7	21			
Turkey[d]	2.5	26			
Luxemburg	0.9	41			
Ireland	0.7	10			
Belgium[d]	0.2	1			

[a] In per cent of total (gross) generation.
[b] 1 Gigawatt hour = 1m kilowatt hours.
[c] 1970.
[d] 1971.
Source: *Annual Bulletin of Electric Energy Statistics for Europe* (New York: United Nations, 1973).

quantities of crude and refined oil that had to be imported by nearly every country, France and Austria imported one-third of their natural gas requirements, West Germany one quarter, and an even smaller proportion was imported by the United Kingdom. Imports were also significant in the case of coal. Because of the declining part played by solid fuel in each country's total primary energy needs the order of magnitude of the imports of individual countries was very different. This ranged from practically the whole of the relatively small Italian or Swedish requirements to only 23 per cent for Spain. Even the once large coal producers had to import some of their coal requirements: about one-half in the case of Belgium, one-third in France, 40 per cent in the Netherlands and 5 per cent in the United Kingdom.

Energy Policy before the Crisis

There has never been such a thing as a "European" energy policy. There has not at any time been a cohesive force which might have coerced the countries of Western Europe into negotiations on this subject. The European Community has shown no signs of positive interest, although the present crisis may push those responsible into seriously considering the matter. This section of the chapter is restricted to a brief survey, necessarily short and selective, of the energy polices of the various countries before the crisis was finally precipitated.

It is no surprise that, even within the European Community, policies were extremely divergent. The nine members of the European Community are still independent sovereign states hitherto following very different lines on energy policy – in so far as they had a consolidated and comprehensive energy policy *per se* (which cannot be said of most of them). Indeed it is very little exaggeration to say that in the nine countries there have been, and are, nine different energy policies.

A main cause of those differences comes from the presence, or absence, of a coal-mining industry. Despite the different degrees of protection given to coal during the postwar decades of cheap and abundant oil supplies – protection which varied from country to country but which was in all cases significant – only two countries in Western Europe now have major coal industries, West Germany and Britain. British production of hard coal is now only marginally smaller than that of the rest of the European Community. Whereas countries without a coal industry had no reason to support coal, those with coal industries struggled with the substantial problems that the gradual contraction of this once powerful sector of the economy produced. These were not only economic and financial but also social, and it was this last factor which was of greatest importance for some countries. Traditional coal-mining, as direct from open-cast or strip-mining, has certain affinities with agriculture, being not just a profession, a vocation or a skill, but a way of life. This led to a curious if understandable contradiction that whereas in certain locations the industry's contraction closed many mines, with consequent redundancies and all the allied worries, in other areas it proved difficult to replace retiring miners by new recruits so that economic pits could be kept going. This required a special employment policy, as well as high earnings for the miners, which resulted in frequent industrial unrest, more severe in some countries than similar disputes in other industries.

Different views have also been taken of oil industry developments. Britain and the Netherlands, being hosts to major international oil companies (notably British Petroleum and Shell), took a view which differed from the attitude of the other countries on such matters as oil supplies, pricing and even the penetration of American oil companies. They tended to take a "world-wide" view of the oil industry, and one that was not entirely free from political overtones. France, on the other hand, was inclined to develop national companies; and her interest in Algerian sources overshadowed her political activity in the oil world. Italy, poorly endowed with indigenous energy sources, has made great efforts to build up a national oil industry relying on all possible sources. The Dutch encouraged the massive concentration of oil refining in Rotterdam with the aim of becoming central refiners for Western Europe. Those European Community members with access to major natural gas resources – the Netherlands, with her huge Groningen field, and Britain with the North Sea finds – were not prepared to hand over these resources to Community control.

Cooperation on nuclear matters was equally unenthusiastic. Britain, starting first, went alone (obviously, not being a member of the European Community); but France and West Germany also developed their smaller nuclear programmes independently of each other. Although everybody now agrees in principle, in view of the massive resource requirements involved, that future research should be undertaken jointly or at least strictly harmonised, the prospects for the immediate future promise no more than very moderate advance in this direction and on restricted lines (such as uranium enrichment), with nothing in other areas. This represents no more than marginal progress on the whole front.

Further, and perhaps more important, differences arise from the basic attitudes to economic management in general. The two extremes are the West German "free market" policy and the French *dirigiste* tradition. To this can be added the varying degree of public ownership of the different energy industries. This results in a wide array of structural and organisational variations with almost endless possibilities for permutation. These naturally influence – mostly hindering and rarely promoting – any attempt to achieve a common energy policy within the Community. They provide the main explanation as to why national decision-makers find it much more difficult to get seriously involved in discussions aimed at a common energy policy than do the Brussels administration.

It is not entirely true to say that the European Community had made no preparations before the crisis. Back in 1968 a document, *First Guidelines for a Common Energy Policy*, was submitted by the Commission, but its impact has been virtually nil, even though it has formed the basis of all subsequent discussions on energy, even post-crisis discussions, that have taken place between member countries. A report in the *Financial Times* of 24 May 1973, only five months before the crisis flared up, was characteristic: "After over 18 hours of marathon negotiations, Common Market ministers were forced early today to accept that their first meeting in over three years to discuss energy policy had been hardly an unqualified success". In fact it was no success at all. Although the underlying problems had been correctly recognised by the Brussels administration, it is obvious that for a long time these were not regarded as sufficiently important to justify serious discussion. Even in mid-1973 they were not thought to be sufficiently pressing for the representatives of member countries to make the effort required to reach that degree of agreement which might have brought a common energy policy nearer.

There has, however, been a general recognition[7] of the need for solidarity as well as a proposal for a common purchasing organisation (with membership open to third countries). Further the wider membership of the Organisation of Economic Cooperation and Development (OECD) accepted a recommendation that every member country should hold stock equalling ninety days consumption of crude oil, and also adopted, in November 1972, an emergency scheme to distribute 10 per cent of available energy products for special allocation. The possibility of future difficulties affecting all member countries, and the desirability of collective action, was recognised, but it was not followed by efficient action; even the implementation of the OECD recommendation and decision left much to be questioned. Whilst earlier attempts for common procedures in the energy area – the European Coal and Steel Community and the European Atomic Energy Commission – undoubtedly achieved partial successes in selected and often isolated spheres, they were not intended to approach the whole complex of energy; consequently the study of energy policy in Western Europe must rely on national endeavours.

The most superficial examination of the energy policy statements of governments in different countries is bound to illustrate the similarity of their general approach: though expressed in a variety of ways, the objectives of each were to satisfy the country's needs and to provide cheap energy. Security was often mentioned, but usually as of secondary importance. None of this, of course, is surprising. Nor is it surprising that "cheap energy" very often meant cheap oil – because if precise account is taken of the various supports given to domestic coal industries, the overall result was rather expensive energy. These supports were given from general taxation revenue in order to keep the price paid by the user low and thus generate the illusion of cheap coal. Similarly if the total investment in research and development leading to nuclear generation is taken into account, electricity from this source – for example in the United Kingdom – is also a rather expensive form of energy.

As long as the buyer's market for oil continued, there was little to fear from the interruption of supplies. Until 1973 Western Europe suffered very little from energy shortages. There were occasional stringencies caused by sudden surges in demand, by strikes in the coal industries, or by political or quasi-political events – such as the Suez incident in 1956, the Arab–Israeli war in 1967, the transport difficulties arising from the closing of the Suez Canal or the repeated blockage of the trans-Arabian pipeline. These were solved either by straightforward imports of (mainly American and Polish) coal or by the flexible and successful operations of the international oil companies. Hence, on the point of security, the record was a good one.

There has been a wide variety of approaches in the way that individual countries have attempted to achieve their policy objectives. The West German economy was for some years overshadowed by the postwar reconstruction. Then during the period of cheap and abundant oil supplies, the coal industry – except for lignite – started to decline swiftly and eventually required consolidation. This aimed at the total reorganisation of the hard coalmining industry (especially the Ruhr). Statutory intervention resulted in the founding of Ruhrkohle AG, a company guaranteed by the federal state, carrying on the business of a large number of mining companies. This was the first really major state intervention in the energy industries of West Germany; although many of the power stations are in public control, they are in municipal rather than state ownership. (As a point of interest: the majority of lignite producers are also owned by these electricity enterprises and supply most of their output to the parent company's power stations.) The state intervened in the oil industry as well, with a grant to a newly formed company, Deminex, for the exploration for oil resources. West Germany's huge refinery capacity is mainly foreign-owned; not more than one-quarter of it is in German hands. The first experimental nuclear power station was commissioned in 1960. By 1973 West Germany had nine stations (2,100 megawatts capacity) and thirteen further plants in course of erection. Nuclear generation is expected to rise from 4 per cent in 1974 to 15 per cent of all generating capacity by the end of 1977.

By contrast to the above, the British energy industries are all in state ownership with the exception of the oil industry. Despite this difference the British coal industry shared the West German troubles, regardless of support, including a tax on fuel oil, and pressure on power stations to burn coal even when it was not the most economic fuel. The successful implementation of the Clean Air Act, in many ways a remarkable achievement, helped to reduce demand for coal. A major political step was the introduction of the first nuclear programme; at the end of 1971 there were nine commercial stations in operation with a capacity of 4,500 megawatts, and a further 8,000 megawatts nuclear capacity of a more developed type is planned for the late 1970s. A larger 250 megawatt prototype fast-breeder reactor is also expected to be normally operating shortly. A welcome addition to indigenous energy supplies, and one which affected policy, was the discovery of considerable natural gas deposits in the North Sea; exploitation was quickly developed in the second half of the 1960s. This changed the gas industry from a stagnating maker of town gas to the reformer and distributor of North Sea gas, which has given gas a significant and growing share of the energy market. A few years later oil was also discovered in the North Sea and the potential from this source, although not very important in world terms, might cover a very large part, perhaps all, of British requirements by the end of this decade. The various companies engaged on the exploration work have been facing a series of technological difficulties due to the adverse conditions of the North Sea, but oil started flowing in June 1975. The North Sea discoveries provided the spur for political decisions designed to minimise any harmful effects of the "oil rush" on the environment of the Scottish northern region, including the islands. The price to be paid by the nationalised gas industry to the gas-producing oil companies was established (where there still appears something to be learnt from the American experience, namely the disadvantages of fixing prices at levels too low to encourage further exploration work). The question of "British" or "state" participation in North Sea operations still has to be resolved (not to speak of specific "Scottish" participation).

The French also nationalised a good part of their energy industries and contained them within the framework of successive national plans. Thermal power stations were regularly supplied with domestic and "Sarre" coal, which was far more costly than imported coal or oil. Nevertheless the French coal industry shared a similar fate to its bigger European neighbours. Competition (often regarded as irrational, if not definitely harmful) among the various energy industries, mostly in public ownership, was minimised through the use of regional policies and various systems of preferences. The French nuclear programme started much later than the British (and with less financial input into research and development), but its further progress is planned to be rapid.

By sharp contrast to the energy policy of these larger countries is that followed by a smaller but highly developed country. Sweden has no coal or oil but is geographically well provided with hydropower potential, which was developed many years ago and became one of the pillars of Swedish industrialisation. Nowadays it supplies a minor part of their primary energy requirements, the rest coming mainly from oil. The Swedes have followed, from the beginning, a policy of free competition and as a result – largely due to the free import of Soviet fuel oil at low prices – they claim to enjoy the cheapest wholesale prices in Western Europe (which supplies most Swedish imports). The energy-producing industries, including hydropower companies, are not nationalised although the State Power Board alone transmits power at voltages exceeding 130 kilovolts. Three recent important lines of development have been (i) the expansion of refinery capacity (it was not until the end of the 1960s that even one-third of domestic requirements could be met from local refineries); (ii) the construction of thermal stations based on oil in order to supplement the hydropower potential; and (iii) the initiation of a nuclear programme. A Swedish nuclear reactor for commercial use went into operation early in 1972 (440 megawatts), the first in a programme which aims to build about 6,500 megawatts of nuclear capacity by 1979.

A more thorough study of the energy policy of these countries, as well as the policies of other countries, reveals that in most countries there is evidence to support the assertion of one observer that in France "a voluntary collusion between top civil servants and managers of large firms"[8] existed in some areas. Coal protection was a particular case in point – although in individual countries to what extent it was collusion or enforced instruction remains debatable.

One further important point becomes clear, which especially (but not exclusively) affects heavy electrical capital goods; namely the impact of the state, as the most important and often the sole purchaser of such products. While not unique, France supplies a good example. The national electricity corporation indirectly implemented a specific industrial policy, by reserving orders to the French industry, by encouraging firms to specialise and merge until they reach the (in France) proposed optimum number of – say – three firms per sector, and by imposing norms and specifications. In a West European context however these attitudes – notably the minimal admission of foreign suppliers for reasons of strategic safety – have resulted in a curious situation. Apart from a larger number of mainly specialist companies in this industry, there are six major groups in five countries,[9] each supplying the full range of electric power equipment. There are doubts whether the market size justifies this number of enterprises and many experts believe that if European suppliers are to match the economies of scale of similar groups in the United States, and increasingly in Japan, then West European concentration, across boundaries,[10] will be necessary.

OUTLOOK FOR ENERGY SUPPLIES

Energy shortages are not new to Europe. One of the first was in Athens at the time of Pericles in the fifth century B.C.[11] when wood was the chief source of energy. Similar worries troubled the people of Elizabethan England in the 16th century when "the beginnings of a timber famine stimulated the search for other fuels"[12] which continued into the 18th century.[13] In this section an attempt will be made first to assess the extent of the present shortage, then to examine the outlook for energy supplies in the short and the medium-long term, before turning to the demand side of the equation in the next section.

In view of the importance of oil – and the significance of the Middle East and North Africa as a supplier – it is convenient to start with the vexed question of oil resources. No doubt the world's oil resources are finite, but nobody knows the exact size of world reserves. Three factors are relevant here: (i) known reserves, (ii) possible deposits in unexplored areas and (iii) the rate of extraction.

Of the three, only the first can be quantified, if only approximately. There are numerous estimates from geologists. All of them suggest that world reserves, at the present rate of extraction, could satisfy demand for several decades to come (at least until the beginning of the next century, and perhaps considerably longer), although they disagree in their estimates of the reserves. It is also agreed that the crude oil output of some of the important producers could be doubled straight away, in exceptional cases perhaps even trebled, without any problem of engineering or geology.

Reserves in unexplored areas obviously present a bigger problem. There are many parts of the world which are still unexplored. Ten years ago few would have expected to find oil under the North Sea. The significant find in Alaska and the more modest deposit discovered in the Aegean Sea were other surprises.

Finally, with present techniques, only a part – generally under a third – of the known deposits is actually being extracted. If recovery rates can be improved, this could mean an immediate addition to the accessible known reserves. Currently a 1 per cent improvement in average efficiency from all known commercial fields would add, it has been estimated, another year's production or more. Whilst to some extent such an improvement depends on improved technology, secondary recovery methods which hitherto have not been economic could now look more attractive when viewed against the higher price of oil.

Reasons for Limiting Oil Extraction

Whilst the position is not as hopeless as some commentators claim, and much depends on future technical advances, the fact remains that whatever is extracted now and in the future comes from an undefined but finite quantity. The oil-producers have made this the focal point of their argument. But they have more practical considerations in mind as well – leaving aside the use of oil as a political weapon. For some of them cannot absorb the earnings stemming from the sale of ever-increasing quantities of crude oil at the new and higher prices. This is an important point which should not be belittled since some of the most important producers, commanding the largest reserves, fall into this category. How strong this argument can be for main producers depends on the size of their population, the requirements for their development plans and the political attitudes of their governments. Another factor is the size of their known (proven) reserves. Obviously the shorter the expected life of the reserves of an oil-producing country the sooner the bonanza comes to an end for them and it is understandable that they would wish to postpone the time of exhaustion.

It is not possible to demonstrate every feature of this phenomenon, but Table 6.6 will suffice to illustrate what is meant. This classifies the chief oil-producing and exporting countries according to the size of their 1973 output, the lifetime of their proven reserves at unchanged 1973 levels of extraction, and the size of their population.

The needs and objectives of a country with a small population, such as Abu Dhabi with less than 0.25m, Libya (2m) or even Saudi Arabia (8m) are quite different from those of a larger or more developed country, such as Iraq, Iran or Algeria. The disturbing fact however is that among the countries with more plentiful reserves Iraq alone appears to have a reasonable absorptive capacity. The others – and chiefly those with large output such as Saudi Arabia, Kuwait and Abu Dhabi – probably do not have much incentive for higher output. Other countries (such as Libya) may be in the same position with the additional disincentive of relatively smaller reserves.

Unless this situation can be changed (by providing some novel incentive), the outlook is not very promising, at least not for excess shipments from many Arab producers. Even among the Arab countries, though, attitudes differ. During the limitations imposed in the winter of 1973–74 Iraq for example did not follow the cut-backs in production and Algeria too took much milder action than most other Arab countries. Non-Arab producers, including Iran, actually increased their production and shipments. Output was also stepped up in those countries which, although

TABLE 6.6
Crude Oil Reserves of OPEC Countries

1972 production in 1,000m barrels	*Lifetime of reserves in years*			
	More than 40	*30–40*	*20–30*	*10–20*
Over 2	Saudi Arabia (8.0)		Iran (30.6)	
1 to 2	Kuwait (0.9)			Venezuela (11.0)
0.5 to 1	Iraq (10.1) United Arab Emirates (Abu Dhabi and Dubai) (0.2)	Libya (2.1)	Nigeria (58.0)	
0.25 to 0.5		Qatar (0.5)	Algeria (15.3) Indonesia (121.6)	
less than 0.2	Ecuador (6.5) Egypt (34.8) Oman (0.7)	Colombia (22.5) Trinidad (1.0)	Tunisia (5.4)	Bahrain (0.2)

Note: Figures in brackets represent population in millions. (The figure for Saudi Arabia is an estimate.)

importers, have domestic oil resources as well. Outstanding among these is the United States. Since supplies for Western Europe depend on the worldwide situation – and since, as explained in Chapter 1, the upsurge of American imports definitely contributed in a marked way to the course of events – it is most important that the measures taken by the American authorities result in the raising of oil production in the United States.

Whatever the developments in supply, it is unlikely that Western Europe can expect a continued unlimited supply of oil, which in the previous decade permitted an annual increase of oil consumption exceeding 10 per cent. If past growth of demand for oil is taken as a yardstick, Western Europe would suffer a shortage even if oil supplies continued to grow at a more moderate rate, let alone remained static or declined. Oil production and deliveries are a function of political as well as economic considerations and therefore it is difficult to see ahead. The likelihood is that in the next few years the same, or a somewhat higher, quantity of oil will be available as in 1973 but at much higher prices than prevailed before October of that year.

Possibilities for New Sources of Supply

In the short term the replacement of oil is a difficult matter. High prices will naturally be a strong incentive for finding substitutes. But there are considerable difficulties to overcome on two grounds: technological feasibility, and availability of other forms of fuel. Very few consumers can switch overnight to the use of another fuel. Power stations equipped with dual firing systems can do this, but except for a few exceptional cases elsewhere that is about all. Many industrial users however can convert their plant to other fuels within a relatively short period, which may be not more than a few months. It requires some investment as well as adequate capacity to produce and install the equipment. It furthermore assumes that some other fuel will be available. Coal is the obvious candidate, especially in countries where there is still a coal industry. Natural gas may also be considered for some applications. No doubt the coal industries in Western Europe will be given a new lease of life. For they have suddenly become economic as compared with the higher price of oil products. The expansion of coal output is therefore possible. Current workings can produce more coal, and it is possible to open up new pits which would produce high quality coal efficiently (such as in Yorkshire), although it would take a number of years before actual production in the new mines could start.

Apart from such possibilities, which on balance are of a minor order, the price-elasticity of demand for oil is, even at present prices, relatively low in the short term. But it is believed that the longer the period, the higher the elasticity of demand becomes – that is, the more the demand for oil will depend on price. This however assumes that price differentials can generate technological advance. The sudden jump of oil prices will have three major effects:

(a) it is a great incentive to economise on the use of oil;

(b) it will intensify the search for new forms of energy and new sources of traditional types of fuel; and

(c) it will improve the economic viability of other sources and forms of energy.

The first of these effects starts to work immediately, although the results may take some time to materialise. It may be an important factor in reducing demand, or keeping it within limits, and this will be discussed in greater detail later in this chapter. The search for new sources of fuel is a different matter, for it is time-consuming, requiring considerable input of often scarce resources (especially money and brainpower); and its outcome is difficult to foresee, since inventive ability and innovative activity are involved where the dividing line between reality and science fiction is rather blurred.

Besides coal, Western Europe can realistically hope for modest additions to its resource base from the North Sea. Commercial considerations (notably price) aside, there are good prospects for increasing the quantity of natural gas from the North Sea, from present producing areas and from new ones, as well as in the form of associated gas from oil wells. The expected minimum crude oil from the British part of the North Sea alone is now estimated to exceed 100m tons a year by about 1980; although some estimates expect a much higher output, even this minimum should cover the bulk of British requirements. The potential Norwegian production will be far in excess of that country's needs. Exploratory activity is intense – in many places not very far from the take-off stage – in others parts of the North Sea. Many other offshore areas around Western Europe are still unexplored, including, for example, the Celtic Sea.

Another question mark hangs over the export potential of the

Soviet Union. There are two questions here: (i) the ability and (ii) the willingness to export oil and natural gas to Western Europe. Soviet reserves are enormous, but mostly in faraway locations such as Siberia. Uninterrupted transportation would require the building of thousands of miles of large-diameter pipelines. Although technically feasible, this assumes tremendous investment. A further question is the organisation of the production in adverse conditions. This again could be solved with the "know-how" which is either available within the Soviet Union, or could be supplied by other countries. Whether the Soviet authorities would wish to expand their sales of energy to the countries of Western Europe, and to what extent these countries would wish to be dependent on them, are questions involving internal and external political issues. Obviously the oil and gas requirements of the Soviet Union and of its East European allies will continue to rise rapidly, and therefore internal needs are likely to receive preference. Nevertheless in exchange for such items as advanced technology, there might be substantial possibilities for securing access to Soviet oil or gas in larger quantities than already received by some West European countries (Finland, Sweden, Italy, Greece and – more recently – West Germany).

Even if all the hopes for newly-found oil and gas and Soviet possibilities are realised, however, Western Europe will still require huge quantities of oil. These can only come from the traditional producers, chiefly the Middle East, unless new sources are further developed or found, either within Western Europe or elsewhere in the world. Pressure on the available and exportable quantities would thereby be eased.

It is in this context that the tar sands and oil shales around the world should be considered. Most of these are in North America and are said to contain enormous amounts of oil – comparable with Middle East quantities. So far, apart from pilot attempts, no major exploitation of oil from tar sands or shales has been seriously attempted on the grounds that the product would not be competitive with conventionally produced oil at the old price. The new price naturally casts the economics of this kind of production in an entirely different light. Feasibility studies have disclosed the magnitude of the problems involved. These are of two kinds: the resources are virtually in a solid state, in remote areas, and require mining and processing operations which would provoke serious environmental questions, apart from needing large resources in manpower, equipment and investment capital. Furthermore in this particular case more is needed than just a calculation indicating the pay-off of the capital invested. Huge amounts of energy are required, at the present state of technology, to recover the oil from the sand or shale. The "energy balance" of the planned operations seems rather unfavourable. It has been suggested that the extraction of one barrel of oil from shales may require very nearly the input of one barrel of oil! Regarding tar sands, the "energy balance" may be marginally better, with three-quarters of a barrel of input for every barrel of output.[14] Such scepticism is not generally held and no doubt the technology of these novel operations can be developed. But at this stage it is too early to attach great hopes to them.

Scope for Replacing Oil

In certain applications oil might be replaced by liquid hydrogen or methanol. Although petrol (gasoline) has many advantages over them in internal combustion engines, there have already been experiments in the United States with hydrogen-powered aircraft. There are also advocates of methanol which, at least according to some reports, can be produced from coal, is cheaper than petrol and is free of the latter's harmful environmental effects[15] (as is hydrogen).

Coal can provide the base material for synthetic crude oil and also natural gas. Again the productive equipment is not available at present and, in order to raise the process to a scale yielding commercially meaningful quantities, new technology is required which is superior to that applied in South Africa. The latter has been in operation since 1955, producing the full range of oil products, and is based on the Fischer-Tropsch process – in use before and during the Second World War, chiefly in Germany. In this direction the results of the large-scale experiments in Scotland, by British Gas, appear promising.

It is quite likely that Western Europe's hydropower potential will again be considered in the light of the very high oil prices. Some projects which, earlier on, could not be considered on economic grounds might prove feasible. Apart from the conventional concept of water power, other – newer – ideas may also be re-examined such as whether or not tidal power can be utilised elsewhere, following the example of the Rance station in France;

and new pumped power stations may be added to those already working (although these do not produce primary energy, they have some part to play in satisfying peak electricity requirements).

Nuclear power will receive a boost. At the old relative costs, many nuclear power stations reached the break-even point in the cost comparison with coal or oil-fired thermal stations, although their contribution to power supplies was small. The main problems stem from the technology as well as from the safety aspects and the disposal of heat and radioactive waste. It is feared that nuclear plants on the Rhine could cause fogs in low lying parts of West Germany. Nevertheless in all countries a significant expansion of nuclear power is planned; indeed the contribution of nuclear power to electricity generation is hoped to rise from 3 per cent in 1974 to nearly 30 per cent in 1985.[16] A possible limiting factor could be the resources required – including in this case the technology – since regardless of the accuracy of the above forward estimate it seems certain that nuclear capacity will grow very rapidly. But it will require considerable amounts of nuclear fuel, basically uranium and thorium. World resources of nuclear fuel depend on price: at a low price they are limited, at higher prices they are much greater and, if low-grade resources are allowed for, they can be very great indeed. In addition there is the higher cost of the as yet undeveloped processing technique of obtaining them from sea-water and if necessary, from ordinary granite. (If the fast breeder reactor operates successfully, the problem becomes much smaller since this reactor produces more fissionable fuel than it consumes.)

There are many other avenues of research and development in the energy area: wind power, solar energy, geothermal power, nuclear fusion, other ways of generating electricity (magnetohydrodynamics, etc.), as well as research in others areas (such as the automobile driven by non-petrol fuel), but most or all of them are at an early stage of development and their practical contribution to energy supplies is impossible to estimate at this stage.

To sum up: whereas in the short term Western Europe cannot escape depending on oil supplies from the Middle East and elsewhere, the outlook for the longer term appears more promising, much depending on technological advance. If some of the many research and development projects yield a commercially viable energy innovation, this dependence could decline very rapidly, if not disappear.

DEMAND FOR ENERGY

International bodies, academic scholars, civil servants and others produced during the early 1970s a respectable number of forecasts concerning the demand for energy at some future date in Western Europe as a whole, in the European Community or in individual countries. It is beyond the scope of this chapter to assess and summarise them. It is enough to say that most forecasts were prepared well before the 1973 crisis and therefore could not take its impact into account. Some of them clearly foresaw the coming difficulties, but expected them to arise some ten to twenty years later. It would have been impossible to allow for the impact of either of the two major aspects of the crisis, the shortage and the price rise, since there was no precedent for either of them, which meant there was no basis for assessing the future course of development. They agree however in one respect. Each of the forecasts expects Western Europe's energy requirements to continue to grow rapidly. It is this point which should be questioned in relation to availability and cost.

Under normal conditions the employment of energy would continue to grow. Experience suggests as much, and many studies, econometric or other, have provided evidence to support the premise that the use of energy has been growing in line with economic growth. This was so in the past. But in future, energy will be less plentiful, and it will be much more expensive. Having briefly surveyed the outlook for supplies, it is now necessary to examine critically whether demand can be reduced or restrained – and, if so, how and where.

Proposals for Saving Energy

The *per capita* energy consumption of West European countries varies extremely widely. This cannot be fully explained by the varying degrees of economic development of these countries nor by climatic differences. The study of the structure of consumption or the structure of industrial production also fails to explain these huge variations between countries which otherwise would be thought to be comparable. A more thorough study of this problem would certainly reveal areas of consumption where extravagant and wasteful usage of energy may become obvious.

More recently the efficient utilisation of energy was the subject of a high-level conference organised by the Science Committee of the North Atlantic Treaty Organisation (NATO).[17] The report of this conference contains proposals for a large number of energy-saving measures. Some are just ideas. Others could be implemented relatively easily. Their starting point is that no matter how justifiable some practices seemed in the past, they may now be too profligate in their use of energy to be tolerated any longer. The simple examples are the widespread wastage of materials – such as glass and paper – in which a great deal of energy had been invested, or the use of inefficient small electric motors in industry. There are also general principles of raw material conservation which, in the view of the experts who produced the report, must now be reconsidered, such as "built-in obsolescence". The report quotes for instance the quantity of energy required to reduce underbody corrosion in a motor car, which is negligible compared with the energy conserved by extending the car's useful lifespan. This suggestion may be extended very widely since it is common knowledge that many consumer goods – a typewriter and so on – are nowadays constructed for a relatively short lifetime. Many of the proposals require considerable further development before implementation. The real value of such reports lies perhaps more in the novelty of the approach; they direct the attention to hitherto neglected areas.

Indeed if consumers in industry and elsewhere will pay the same attention, under the pressure of high prices, to "fuel productivity" as they used to pay to labour productivity, considerable savings of energy could be reached. The past cheapness of energy, and especially of oil, was a disincentive in this direction and a certain negligence and wastefulness characterised the energy economy of many if not all consumers in all walks of life.

There are many examples – some of which have been reported in the press – of how up to 5 or 10 per cent of the energy consumption of industrial plants could be saved, without any detrimental effect on output, by modest organisational changes. There are also examples showing that if such a reorganisation is on a larger scale, and some investment capital is spent on new power equipment, the saving can reach 20 or 30 per cent. At present energy prices the pay-off period of any such investment would be short.

An important aspect of energy-saving is to survey critically whether the various types of energy are used in appropriate applications. The best instances of obviously wasteful usage are electric space-heating or the use of natural gas, a "premium" fuel, for steam-raising, where coal or residual fuel oil could perform the same service. Of course any such generalisation is dangerous: it would similarly be wasteful to go back to coal on the railways where (at least in Britain) one ton of diesel oil replaces at least five tons of coal.

The national and the individual interest may not coincide here, but if all types of energy were used for exactly the purpose they serve best, a country's primary energy bill could be reduced, often significantly.

Other areas where research, development and implementation of a concerted "saving" policy appear very promising include:

Heat-recycling: In industry especially, heat could be used again rather than dissipated.

Insulation: British buildings for example require more than twice the quantity of energy for heating than similar buildings in Scandinavia, due solely to inferior standards of insulation.

District heating schemes: These are much more advanced in some countries than in others, although they too have their problems.

Increasing efficiency of power generation: It is well known that not more than a fraction of the energy output is utilised in present systems of electricity generation.

Transportation of energy: At present, as Table 6·7 shows, losses in transport and distribution are very great; any reduction in these areas would immediately result in large savings of primary energy requirements.[18]

Recycling of waste: This could take the form of recycling heat produced in energy transformation, which at present is mostly wasted, as well as the recycling of waste materials. Great possibilities exist in both areas, although caution is needed since the recycling and upgrading of certain waste materials may require more energy than the production of the natural material.

Alternative technologies: Possibilities exist for switching technologies towards those processes utilising less energy. Knitting for instance requires less energy than weaving and in many applications the two processes are competitive. Similarly laser and electron beams consume less energy than the processes generally used to welding, and so on.

TABLE 6.7
Losses in Transport and Distribution, 1972

	Electricity[a]	*Gas*[b]		*Electricity*[a]	*Gas*[b]
Austria	7.8	6.2	Netherlands	5.7	. .
Belgium[c]	5.5	1.7	Norway	9.3	4.6
Denmark	10.5	3.3	Portugal	12.8	2.2
Finland	7.1	. .	Spain[c]	13.1	2.3
France[c]	7.1	. .	Sweden	12.1	. .
Germany (FR)	5.2	0.2	Switzerland	9.6	9.9
Greece	8.1	0.6	Turkey[c]	19.4	11.3
Ireland	10.6	. .	United Kingdom	7.7	7.1
Italy	8.3	3.5	Yugoslavia	12.3	. .

[a] Reported losses as per cent of supplies available for consumption.
[b] Reported losses as per cent of deliveries for final consumption.
[c] 1971.
Note: The variations between countries may be partly due to different methods of measurement.
Sources: *Annual Bulletin of Electric Energy Statistics for Europe* (New York: United Nations, 1973). *Annual Bulletin of Gas Statistics for Europe* (New York: United Nations, 1973).

"*Self-generation*": In certain applications, industrial users may find it more economical to generate electricity themselves, instead of receiving it from the national grid system, because of grid losses; "self-generation" is much more popular (Table 6.8) in some countries than in others, though this again is one of the areas where national and individual interests may clash.

Location of power stations: The new energy prices may cast doubt on past practice favouring the concentration of power generation, partly because of the high transportation costs of electric power and also because of the high transmission losses. When bulky coal was the basis of electricity generation it was understandable that it was easier and probably cheaper to transport power than coal. But nuclear fuel, which is relatively easy to transport and in quantities that will be small, is beginning to play a bigger role; further, some power stations favour residual fuel oil which is also easier to transport than coal. Consequently the costs and benefits of producing electricity at giant stations and transmitting the power should again be carefully compared with a less concentrated generating system (still using large equipment, but not in sets of four or six) since the reduction

TABLE 6.8
Electricity: Public and "Self"-Generation, 1972

	Per cent in total			*Per cent in total*	
	Public	*"Self"*		*Public*	*"Self"*
Austria	83.0	17.0	Netherlands	86.7	13.3
Denmark	97.6	2.4	Portugal	93.2	6.8
France[a]	79.5	20.5	Spain[a]	95.1	4.9
Germany (FR)	71.5	28.5	Sweden	85.8	14.2
Greece	96.9	3.1	Turkey[a]	87.8	12.2
Ireland	97.1	2.9	United Kingdom	91.9	8.1
Italy[a]	78.4	21.6	Yugoslavia	94.3	5.7

[a] 1971.
Source: *Annual Bulletin of Electric Energy Statistics for Europe* (New York: United Nations, 1973).

of transmission losses may easily outweigh other advantages. This however is a very complex question and the foregoing is a somewhat over-simplified exposition of a part only of many allied problems.

Automotive use of energy: Although too much emphasis is frequently placed on the importance of the automotive sector in discussion on energy, and especially oil, this area is unique in that the oil products used at present cannot be significantly substituted by other energy sources. The energy requirements of public transport are naturally much smaller on any relative basis than those of private transport,[19] and therefore the transport policy of many countries (especially where public transport has been reduced by dismantling some facilities and reducing the capacity of the remaining networks) will have to be reconsidered. Goods transport and passenger transport both fall in the same category: the "energy-efficiency" of truck transport compared with the goods train is just as low as that of the automobile compared with the passenger train.

Revision of tariffs: The present tariffs for electricity or gas for domestic consumption do not encourage energy-saving – if anything they discourage it. By charging reduced rates for progressively rising consumption, the more the consumer uses the lower his unit cost of electricity and gas, and even of coal and oil, will be. The London tariffs, as shown in Table 6.9, provide a good

TABLE 6.9
Typical Retail Prices of Electricity and Gas Paid in December 1972 by London Domestic Users

Electricity		*Gas*	
Annual level of consumption kWh	*Price (pence per kWh)*	*Annual level of consumption therms*	*Price (pence per therm)*
750	2.112	80	20.76
2,500	1.301	250	14.76
5,000	1.123	400	12.96
10,000	0.771	800	10.15
30,000	0.704	1200	9.22

Source: *UK Energy Statistics*, 1973 (London: HM Stationery Office, 1973).

example, and there are certainly other places outside the United Kingdom where tariffs similarly discourage energy saving. Such tariffs were understandable when the four fuels were in sharp competition with each other, but their reappraisal is now due in the interests of energy economy.

No such list can be complete, but the above indicates that much can be done on the demand side to reduce energy requirements, without any major effect on the economy or on the way of life, in West European countries.

ENERGY POLICY FOR THE EUROPEAN COMMUNITY

After the jump in oil prices the countries of Western Europe naturally had to take both political and economic measures to contain the new situation. The Netherlands alone was blacklisted in Europe for her friendly attitude to Israel (and Portugal, on other grounds), whereas Britain and France were promised to be treated as "friends" of the Arabs. They therefore could expect preferential treatment in terms of unlimited oil supplies. In actual fact however the international oil companies, fulfilling their normal role of distributors of oil supplies, introduced a system of allocation. As a result each European country obtained, during the critical winter months, relatively comparable quantities. The Netherlands probably did not suffer unduly heavily, and

Britain and France did not benefit too much, from this discrimination – further evidence, supplementing previous experience, that a selective embargo is unlikely to be very effective.

Most countries introduced measures aiming at saving oil, that is, at reduced consumption. As far as was possible, coal started to replace oil, but the scope for this is very limited in the short term. Private motorists' usage was curtailed in a number of forms – by banning Sunday traffic, imposing lower speed limits, and even making advance preparations for rationing. Actual rationing was only introduced in two countries – Sweden and the Netherlands – and abandoned after a few weeks. Electricity use was also curtailed in many countries, affecting advertising and street lights as well as heating. Industries using oil as a basic material as distinct from fuel were also hit by supply shortages. On the whole however major calamities did not occur.

The European Community put the short-term situation on its agenda and suggested, in a proposal for a Council recommendation in January 1974, the harmonisation of voluntary measures for reducing energy consumption. Some of these recommendations were practicable and sensible, such as the reduction of the temperature in factories, offices and workshops by approximately 3°C, or the limitation of lighting and the use of gas and electricity for heating. Some of the other recommendations were hardly applicable to the short-term period that they were supposed to cover, which ended in June 1974. It is difficult in the short term to arrange that people should use public transport in preference to private vehicles (because this assumes an efficient public transport system, which no longer exists everywhere), or to rationalise the carriage of goods by channelling them from road to rail or to inland waterways.

There has been a considerable revival of interest in research and development in energy – one example being the allocation in Germany by the Federal Government of 1,500m marks to energy research projects, on top of the research fund serving nuclear developments. (No other similarly large-scale venture into energy research has so far been publicised.)

Two important international developments were the Washington conference, in February 1974, and the move of the European Community, in March 1974, aimed at contacting the Arab countries as one body and not individually, as some of the member countries had already done. The Washington conference, ended by recognising the advantages of a common approach to the producing countries by the industrial consumers (North America, Japan and Western Europe) which was accepted by eight members of the Community, but not by France, who disagreed with the most significant points of the *communiqué*. In the following month the Community's member countries decided to begin a wide-ranging new dialogue with the Arab countries – wide-ranging in that they covered not only energy but also long-term economic cooperation in others areas, including industry, agriculture, technology "and even culture".

Possible Approaches to a European Energy Policy

The basis of any discussion concerning a common energy policy for the Community is still the *First Guidelines for a Common Energy Policy*, first written in 1968, but updated several times since.[20] It is quite obvious that the enormous changes in relative prices have put some of the recommendations of these "guidelines" in very different light and that, for this reason alone, a major revision of this document is necessary. European energy policy should aim to reorganise energy requirements on a rational basis, maximise forthcoming supplies, and coordinate these two aims. It is perhaps too sanguine to expect an absolutely uniform energy policy to be introduced in all the nine member countries, in view of their individual energy resources and different attitudes to energy questions, and to the steering of the economy in more general terms. These however relate more to the question of the implementation of some commonly agreed principles. Indeed the most likely approach to a common energy policy would seem to be an agreement on the broad outline by the Community, and implementation by the national governments; and it cannot be otherwise, since the executive power lies with the latter.

Some points however may require special emphasis in the light of events since October 1973. First and foremost is the decision on concerted action. The initiation of a dialogue with the producing countries – and possibly with the whole of OPEC and not just its Arab members – is a welcome change in this context but still leaves a number of questions open. Will it lead to some kind of blanket agreement with the producers of oil, covering (within the wide area envisaged by the Community) all transactions of the oil business between the producers and all members of the Community? Or will it still leave the door open for bilateral deals? If it is the former – how can agreements concluded already between Community countries and some of the oil producers be accommodated within it? If it is accepted that some breach of the common front should be – or has to be – admissible, there is still a possibility of some agreed platform for such bilateral agreements which would make it possible to dovetail any such direct deals into a Community approach.

A second important point is the conservation of energy which might better be called the raising of "energy productivity". As already explained, it is in this area that the most immediate results can be achieved for the reduction of import requirements. Achieving this will require additional research and development, which in any case will be of outstanding importance if Western Europe's energy problems are to be solved in the future.

It is suggested that research and development in the area of energy-saving directly serves the short-term future (remembering that in energy matters the short term may mean a number of years), whereas the same activity aimed at the search for new resources and developing the existing ones is of a long-term character. Major programmes in both areas could be very costly indeed and, therefore, it might be sensible to pool the Community's resources, in order to achieve the best research and development results. This does not necessarily mean geographical concentration, but rather a harmonised research and development activity, commonly financed and organised, to coordinate the existing research and development programmes of the various member countries. The outcome of this would not only be joint efforts and hopefully some success in the many obvious avenues open to research and development, which could hardly be covered by the individual countries if they continued fragmented work, but also some kind of centrally operated advisory service which would make the results of research and development available to all member countries in the form of recommendations.

On the supply side such an approach should include (i) an agreement with the oil-producers over maximum continuous supplies, (ii) the best exploitation of existing resources, (iii) the exploration of new resources, not only in traditional but also in new and technologically more advanced lines. Such a policy should be supported by a system of sensible incentives to provide the right "climate" for such developments as are necessary if Western Europe's energy requirements are to be covered. These would include the most efficient utilisation of known resources and the introduction of major innovations.

In reassessing their energy policy, however, the Community, and the individual member countries, will also need to take a hard look at their trade policies. As an instance they could well decide that past policies of matching imports of goods, for example textiles, imported from countries with low labour costs, by the application of advanced but energy-intensive technology may no longer be appropriate under circumstances of high energy costs and associated higher import costs.

It needs to be recognised, however, that the arguments that favour coordinated Community action as opposed to independent action by individual countries apply with equal force in a wider, worldwide, context. The impact of high oil prices, the overhanging doubts about available energy supplies, and more particularly the disturbance to the international system of trade and payments, are matters which affect all countries to a lesser or greater extent.

Any action that the industrialised countries of the Community might take which was much at variance with the policies of the United States and Japan could only be self-defeating. The successful absorption of the surplus funds of the OPEC countries must depend on the ability of the Council of Twenty to negotiate a reform of the international monetary system.

Similarly the advantages that can be derived from coordinated European research and development effort into energy saving would be greatly increased if such effort were coordinated globally, for all consuming countries have similar objectives. It has to be remembered that in basic science a perfectly internationalised system exists within which knowledge and "know-how" move at minimum cost, and an internationally coordinated research effort into energy-saving – and indeed one to increase energy potential – could be achieved with relative ease.

Further, any attempt by the Community for preferential treatment from the producing countries for oil supplies could only be to the detriment of other OPEC customers, more particularly the developing countries, The broad aims of any West European energy policy must therefore take account of the fact that, though imported oil is vital to the continued well-being of the members of the Community, it is no less essential to the economic life of many other nations.

CONCLUDING THOUGHTS

It has already been mentioned that this is certainly not the first energy crisis in European history. But it is admittedly much the most serious in view of the vastly increased importance of energy in present-day life. It is understandable therefore that many Europeans find the present situation rather gloomy. Some of their concern may be eased by considering the immediate situation in a wider time-scale.

In his great work on *Technology and Economic Development*, Asa Briggs describes another gloomy point in the long time-span of European history. In the summer of 1454, one year after the fall of Constantinople to the Turks, Enea Silvio Piccolomini, later Pope Pius II and one of the best informed persons of his time, wrote that he could not see "anything good in the prospects" for Europe. Yet within fifty years Europeans had advanced far beyond the boundaries of their oil continent – to Africa, India and the other side of the Atlantic. And less than 100 years later the great new movements of the Renaissance and the Reformation started to transform European thinking, opening up entirely new horizons and filling life with new content.[21]

The future of Europe's energy supplies apparently relies heavily on new technology. Many are suspicious as a consequence and consider the hopes attached to technical advance over-stated, unfounded and unrealistic. They may be right. But history shows that whilst there have always been those who have foreseen disaster sufficient to equal those who promise success, the doomsters have consistently been proved wrong. This is largely because they view technology as something static and not as the dynamic tool it really is. Even those most closely involved frequently lack the vision to appreciate the full potential of the technology that is in their hands.

Simple illustrations can be given. (i) A leading spokesman of the British steel industry said publicly in 1948 that the application of oxygen in steel-making was an impossible proposition, a non-starter; yet, within five years, this process was commercially introduced in Austria and became the leading new process of steel-making within another decade. (ii) At one time or another, IBM dropped the idea of an electronic calculating machine, and Kodak the concept of what has later come to be called the Polaroid camera. (iii) In the field of jet aircraft, all the great

American names of the industry (Boeing, Douglas and Lockheed) presented massive theoretical evidence indicating that such planes could never become economic, that there was no demand for such high speed, and that flying above 5–6,000 metres would not be feasible for normal passengers.[22]

Whatever success may be achieved in the short term in conserving energy and/or increasing supplies, concern about energy will remain and will continue to present great problems for international trade and monetary policies. But, in addition to the problems, the energy crisis has created an entirely new situation in which there could well be opportunities for moving towards the general integration of the world economy.

The following quotation from a recent report is perhaps an appropriate conclusion to this paper:[23]

> So far man has been saved from exhausting world energy resources by the use of his ingenuity. Perhaps a more basic question than whether we shall run out of the resources we happen to know about at present, is whether we shall run out of the ingenuity necessary to find new forms of energy.

NOTES AND REFERENCES

1. This chapter draws together the material specifically relating to Western Europe in George F. Ray, *Western Europe and the Energy Crisis*, Thames Essay no. 6 (London: Trade Policy Research Centre, 1975).
2. For an account of the developments which precipitated the oil crisis, see Frank McFadzean, "Economic Implications of the Energy Crisis", in Hugh Corbet and Robert Jackson (eds), *In Search of a New World Economic Order* (London: Croom Helm, for the Trade Policy Research Centre, 1974).
3. Christopher Tugendhat and Adrian Hamilton, *Oil: the Biggest Business* (London: Eyre Methuen, 1975).
4. *World Energy Supplies* (New York: United Nations, 1973).
5. Throughout this chapter Western Europe should be taken to include Yugoslavia, as in the case of United Nations statistics.
6. On a *per capita* basis energy demand rose 4 per cent.
7. *On Measures to Attenuate the Effects of Difficulties Inherent in Hydrocarbon Supplies* (Strasbourg: European Parliament, 1973).
8. Andrew Shonfield, *Modern Capitalism* (London: Oxford University Press, 1965).
9. The six groups are: (i) the merged interests of Siemens and AEG in Germany; (ii) the Brown–Boveri group in Switzerland; (iii) the GCE–Alsthom group in France; (iv) ASEA in Sweden; (v) GEC and Reyrolle-Parsons in the United Kingdom.

10. Barbara Epstein, *Politics of Trade in Power Plant: Impact of Public Procurement* (London: Trade Policy Research Centre, 1971). Also see A. J. Surrey, "Towards a Common Market for Heavy Electrical Plant", *The Business Economist*, Oxford, Spring 1973; and Surrey, *The World Market for Electric Power Equipment* (Falmer: Science Policy Unit, University of Sussex, 1973).
11. Wilfred Beckerman, "Economists, Scientists, and Environmental Catastrophe", Inaugural lecture delivered at University Colege, London, 24 May 1972, *Oxford Economic Papers*, November 1972.
12. Christopher Hill, *Reformation to Industrial Revolution* (London: Penguin-Pelican, 1969) p. 88.
13. Max Weber, *General Economic History* (London: Allen & Unwin, 1923).
14. P. F. Connolly, "The Energy Equation", *The Business Economist*, Spring 1974.
15. N. Valery, "The Energy Equation – the Best Substitute for Petrol may be Petrol", *New Scientist*, 24 January 1974.
16. "Electricity in the Mid-Eighties", *ECE News*, Geneva, September 1973.
17. *Technology of Efficient Energy Utilisation*, Report of the NATO Science Committee (Brussels: NATO Secretariat, 1974).
18. *Energy for the Future* (London: Institute of Fuel, 1973). Cost comparisons for energy transport and transmission show the following comparable ratios for conveying unit quantity of energy to unit distance, taking into account running costs, capital cost and depreciation (but not the very high urban land costs):

Oil pipeline	1
Gas pipeline	2.5
Coal train	5
500 kilovolt transmission line	17

19. *Financial Times*, London, 20 November 1973. United States estimates show the following comparison for passenger transport, in terms of passenger miles per gallon:

Large jet plane	22
Automobile	32
Cross-country train	80
Commuter train	100
Large bus	125
Suburban train	200

20. For instance the policy proposals were elaborated upon in April 1974, when recommendations were announced aiming at reducing Community dependence on oil, building substantial nuclear capacity and raising the level of coal usage. But the proposals had a long way to go before they could be implemented. (See *Financial Times*, 26 April 1974).
21. Asa Briggs, *Technology and Economic Development* (London: Penguin-Pelican, 1956) p. 16.
22. O. Stewart, *Aviation: The Creative Ideas* (London: Faber, 1966) p. 205.
23. Colin Robinson and Elizabeth M. Crook, "Is There a World Energy Crisis?", *National Westminster Bank Quarterly Review*, London, May 1973.

CHAPTER 7

Capital Requirements for Developing Alternative Sources of Energy

T. M. RYBCZYNSKI

Following the review in the previous chapter of the impact of the rise in oil prices on the energy scene in Western Europe, an attempt is made in this chapter to evaluate the repercussions of the change in the energy situation on capital formation in the major industrialised countries, by examining the new policies they are adopting and what they imply for financing requirements.[1]

The new energy policies being evolved all round the world – which are likely to be among the most important elements in the overall policies of most governments for several years to come – can be expected to pose profound economic, political and diplomatic issues. How those issues will be resolved is not easy to foresee, particularly in the context of the European Community, as George Ray has shown. Nor is it easy to foresee the outcome in relations between the European Community, the United States and Japan. Even so, the outlines of possible solutions appear to be reasonably clear. In as much as some of them have been incorporated, implicitly, in the analysis used in this discussion, they represent a value judgement by the present writer on the most likely course of events.

MAIN ELEMENTS IN THE NEW ENERGY POLICIES

As a result of the sudden and sharp increase in the price of oil, accompanied by restrictions on its availability, all the industrialised countries have set about evolving new "energy" policies, the main purposes of which are

(a) to reduce the rate of growth of demand for energy,

(b) to reduce the degree of dependence on imported energy by developing indigenous sources as rapidly as possible, and

(c) to ensure the supply of imported oil at reasonable

prices during the transitional period – that is, when the degree of self-sufficiency is being increased to the maximum – and later on to the extent that it will be required.

The relative importance of these three elements – and therefore their probable consequences – differs from one country to another, depending on the existing degree of dependence on imported oil as a source of energy, on the potential indigenous supplies of fossil fuels and on technological and industrial capacity to develop new sources.

For these reasons it is preferable to analyse separately the situations in the three main industrial areas: the United States, Japan and the European Community. While the Community is endeavouring to move towards an economic and monetary union, and in doing so is also trying to develop a common energy policy, the process is taking a long time; and the United Kingdom's position, because of its access to North Sea oil, deserves separate treatment.[2]

TABLE 7.1

Relative Importance of Imported and Indigenous Energy in Total Primary Energy Requirements in 1971 and Potential Sources of Indigenous Energy[a]

	Imported energy as % of primary energy requirements	*Indigenous reserves of fossil fuels*
United States	3[b]	coal, gas, oil – very large
Japan	75	coal – very small
United Kingdom	46	oil, gas and coal – large
France	68	coal – very large; oil – very small
Germany (FR)	67	coal – large; oil – very small
Italy	88	oil and gas – very small

[a] All of the countries listed in this table have the technical capacity to develop potential energy sources.
[b] Excluding imports from Canada.

While policies may differ from country to country, each must take, as one of the principle starting points, the existing and prospective price of oil, though it is probable that their individual judgements as to the likely trend of oil prices will vary considerably.

Assumptions

For the purposes of this paper it has also been necessary to make certain assumptions about the movement of oil prices. These are:

(a) that no significant increase in (real) price levels will take place above those existing at the beginning of 1975, around $11 to $12 per barrel, landed price in Europe, but rather

(b) that there will be a reduction over the rest of the decade, so that by the 1980s the price of oil may be reduced by as much as one-third of the early 1975 levels, to between $8 and $10 an "average" barrel, c.i.f. (that is, prices including cost, insurance and freight) in Western Europe.

The main reason underlying these assumptions – quoted in terms of 1974 real prices – is that the "cartel" of the oil-exporting countries, because of international pressures, will behave in a responsible and co-operative manner.

Until 1970 the international price of oil could be said to have been determined by the marginal cost of oil in the United States – which until then possessed sufficient "marginal" productive capacity to offset any threat of supply curtailment by oil-exporting nations. 1970 saw a fundamental change, for the "excess capacity" in the United States disappeared and the Middle East became the only area with significant surplus availability of oil. This transformed the oil-pricing process from one based on marginal cost to one based on the joint exercise of bargaining power.

Both history and economic theory suggest that the oil surplus countries will continue to have this power – which can be used as an economic as well as a political weapon – as long as there are no alternative competitive sources of supply. It seems unlikely that these could be available before the middle of the 1980s. Consequently any forecast of the world oil price has to be based on a judgement of the future behaviour of the members of the oil cartel in political rather than economic terms.

In formulating the pricing assumptions, considerable weight has been given to the political importance of Saudi Arabia, its interest in the political stability of the Middle East, and indeed of the whole Western world, and its (and other smaller countries') limited absorptive capacity. The Organisation of Petroleum Exporting Countries (OPEC) could well be under pressure from certain of its members to agree on prices which fall short of "what the market could bear". Nevertheless such prices could be conducive to both the long-term progress and the economic and political stability on which their prosperity will depend, at least until the turn of the century.

In the longer run the price used is likely to be in line with the marginal cost of producing energy in industrial countries, thus restoring in a special way the demand–supply situation and enabling the OPEC countries to obtain the benefits of high prices over a very long period. The view that enlightened counsels will prevail during the transitional period may of course be unduly optimistic, and political and other pressures within the OPEC countries may well result in a higher real price than that assumed in the main body of the chapter.

In as much as this occurs the pressure for the accelerated development of indigenous energy resources by the industrial countries will inevitably be greater, the real cost of energy so produced being more favourable than it would be otherwise.

The second major assumption is that the post-crisis goals of the industrial countries' new energy policies will be achieved, even though they may be too ambitious, requiring modification – as happened, for example, in the United States.[3] Some subjective evaluations have been made by officials of some governments, by students of the energy industries, and also by certain industrialists who will be required to expand their capacity and develop new techniques on a large scale. These studies point to substantial delays, sufficient to set back the target dates by some five years – say up to 1985 as compared with the original date of 1980.

In the first instance, consider nuclear power: even with the existing nuclear reactors the construction period is around six years, while for the adoption of new types, as contemplated in the United Kingdom and possibly in the countries on the continent of Europe, the time required will be substantially greater. It is for this reason that the Commission of the European Community appears to have had second thoughts on the feasibility of achieving the revised target of increasing the contribution of nuclear energy to overall energy consumption of 17.5 per cent in 1985, as compared with its original plan of 9.6 per cent. And the United States Federal Energy Office also considers that the application of nuclear energy will only start to become generalised in the period 1985–2000.

As regards production of oil from non-classical sources (oil-shale, oil sands or coal), though basic techniques are known they have to be improved to reduce costs and adapted for large-scale operations. This also applies to gasification of coal and open-cast coal production where environmental considerations now play a very important role. The difficulties so arising are not only likely to extend the time needed to achieve the proposed targets, but may also raise the capital requirement.

In as much as the pressures created will require additional resources, their real cost will be increased. Evidence now becoming available by way of orders in relation to capacity in the United States appears to point in this direction.

POSITION IN THE UNITED STATES

The United States energy policy through the 1960s and up to the autumn of 1973 was intended to maximise domestic production of oil and keep the country's dependence on imported energy at a reasonable level at a time when imported oil was considerably cheaper than indigenous oil. To achieve this aim mandatory import quotas for crude oil (and refined products) were in force, which physically limited the level of oil imports.

The quota system, which involved different methods of calculating quotas for the west coast of the United States and the rest of the country, became increasingly complex over the years of its life, with certain exemptions and exceptions being introduced, making the system more and more unwieldy and discriminatory. During the early 1970s domestic crude production reached its peak and the growing demand for oil had to be substantially met from imports. The increase in demand for oil products was further encouraged by the substitution of low-sulphur oil for high-sulphur domestic coal to meet environmental regulations. This growth in imports caused the authorities to abandon their policy of mandatory import quotas and to replace it with a tariff, thus according the necessary degree of protection to domestic oil-producers while relying on the pricing mechanism to control imports.

Under this system, introduced in April 1973, imported crude oil is subject to a tariff – euphemistically called a "licence fee" – of 21¢ per barrel, while refined products are now subjected to a levy of 42¢ per barrel. Existing quotas are to be phased out over a period of years, and there is a subsidy in the form of free crude imports during the first five years of operation of new refining capacity, equivalent to 15¾¢ per barrel, giving domestic refiners cost advantages of 57¾¢ per barrel, as compared with imported products.

There were four important features associated with, and indeed arising from the policy of mandatory quotas, which are of relevance and importance in the present context. First, despite high domestic as compared with market prices in the rest of the world (a difference which could be as much as $1.50 per barrel – approximately equivalent to the cost of other forms of energy) the demand for energy expanded during the 1950s and 1960s at 4.5 per cent per annum. This rate was as high as the rate of increase in total real output, and in relative terms appreciably higher than in other industrial countries. This relatively high growth rate can be attributed to the fact that, unlike the European countries, taxes (duties) on the final products have consisted of small taxes, not Federal taxes, making the cost of energy to consumers relatively cheaper than in other industrial countries.

The second feature is the comparatively small importance of energy costs in the United States. Estimates available, which take full account of the cost of energy in the intermediate products indicate that in the late 1960s energy accounted for only 2 per cent of total cost of all goods and services, whilst corresponding figures for the Western European countries and Japan ranged between 3 and 6 per cent.

Thirdly, while until the second half of the 1960s investment in domestic energy sources and associated facilities such as transport both in the United States and Canada, was held at the level that would ensure an approximate degree of self-sufficiency, in the late 1960s an increasing volume of investment to cater for domestic requirements was undertaken overseas.

Finally, while until the mid-sixties or so energy did not figure to any significant extent in the United States' external accounts, growing oil imports began to impose a rapidly increasing burden in the late 1960s and early 1970s.

New Energy Policy

Following the steep rises in the prices of oil exported by the OPEC and other countries (starting in October 1973), the US administration announced the basic outlines of a new energy policy in the latter half of January 1974. Its two principal aims were to make the United States virtually self-sufficient as regards energy supplies by 1980, and to reduce the rate of growth in demand from 4.5 per cent per annum to around 3 per cent per annum.

The self-sufficiency is to be attained by allowing the price of energy to rise, thus making it profitable to expand domestic sources of energy, by providing governmental help by way of removing legal and other obstacles; and by providing fiscal and other assistance wherever and whenever necessary.

The reduction in the rate of growth in demand is to be achieved by relying principally on the pricing mechanism (that is to say, by allowing higher prices to restrict demand) together with a number of supporting measures such as maintaining speed limits and altering some of the anti-trust provisions.

Whilst it is recognised that self-sufficiency by 1980 is perhaps too much to expect, the policy could lead to a level of imports by, say 1985, sufficiently low that their interruption for political or other reasons would not be a national calamity. What needs stressing here is that higher domestic prices for oil and gas will make exploration and exploitation of domestic resources, including those in Alaska, very attractive, and will encourage a rapid and substantial increase in the use of coal for gasification, in the exploitation of shale oil, and in the further intensive development of nuclear fission. In all these areas the United States possesses adequate indigenous resources and the basic technological capacity needed to exploit the traditional sources as well as the new ones.

At the same time – and this is of utmost importance – the cost of energy to be obtained from the existing and new sources is estimated not to exceed the equivalent of $6.50 per barrel, as compared with the early 1975 (landed) price of imported oil of around $12.00 and the pre-October 1973 price of some $3.45 per barrel (c.i.f.).

In broad terms the emphasis of the new policy is on increasing, during the next eight years or so, the degree of reliance on coal and nuclear fission, whilst stepping up domestic production of oil – including that obtained from shale oil – and gas, at least to the level of domestic consumption of these products in 1973–74. In essence this would imply a ten- to thirteen-fold expansion in nuclear capacity and quadrupling the production of coal, with more than half being used for gasification.

TABLE 7.2
Sources of Primary Energy in the United States

	1970–72	*1980–82*
Oil	44	31
Gas	33	18
Coal	18	36
Hydro	4	6
Nuclear	1	9
TOTAL	100	100
Absolute equivalent: (million barrels of oil per day)	15.4	21–22[a]

[a] Rates of growth assumed 3 per cent and 4 per cent annual averages.

The new energy policy is formidable and probably the most challenging in America's history, and the precise means by which it will be executed are still far from clear. Remembering however the way the United States created – within two years during the Second World War – the synthetic rubber industry and the atomic bomb, and then the space programme in the 1960s, it can be expected that substantial progress is likely to be made during the second half of the 1970s.

Implications of the New Policy

What are the implications of this policy regarding costs, the rate of growth of the American economy, and the behaviour of its main components, as well as external accounts and trade policy?

The consequences the new energy policy can be expected to have on the behaviour of the whole economy and its main components can best be seen by evaluating the effects of the additional spending. It is estimated that the capital cost of expanding the production of indigenous fossil fuel, of nuclear energy and of ancillary facilities (such as transport and coal gasification) is likely to amount to between $1,000,000m and $1,250,000m or to between $150,000m and $200,000m per annum. (This is of approximately the same order of magnitude as the space programme.) Although the federal government can be expected to provide some help – directly by providing funds and indirectly by giving special fiscal concessions – the bulk of this expenditure will probably have to be borne by private industry.

During 1972 and 1974 fixed capital expenditure by private industry amounted (in 1974 dollars) to some $150,000m per annum, representing approximately 10 per cent of total spending. The new energy policy would thus require a doubling of private non-residential fixed investment in absolute terms, as well as the doubling of the share of private productive capital spending in 1974–75. In the later years, say from 1976 to 1977, the absolute and relative increase would gradually decline as total spending increased, but the rate of expansion would still be very substantial.

What is of greater significance is that additional capital expenditure arising from the new energy policy of $150,000m–200,000m per annum is about 2.5–3.5 times as great as the growth potential of the American economy, which was estimated at around 4 per cent per annum and equivalent in 1974 prices to $55,000m–65,000m. Although acceleration in technological advance by increasing the rate of growth of productivity may well also raise the potential rate of growth of productivity, it is unlikely to alter the picture presented above to any significant extent. Thus the new energy policy will require, and indeed cause, the authorities to endeavour to restrict other types of spending, and above all private consumption expenditures and public sector spending, keeping it very close to its mid-1970s level in the years ahead.

The two broad consequences of the new energy policy are that the American economy will probably experience one of the most intensive and prolonged upswings in its history, and that the requirements of the private sector for outside finance can be anticipated to double, placing a corresponding pressure on interest rates.

As regards external accounts, the new energy policy means that payments for imported fuels, which amounted in 1974 to some $25,000m, can be expected gradually to decline, thus strengthening the United States balance of payments. Implicit in the new policy is also protection for the domestic producers, which would ensure that the price of imported energy does not fall below the equivalent of $6 per barrel. While at present this may be of academic interest, it may early in the 1980s necessitate an increase in tariffs to keep the cost of imported energy in excess of domestic cost, should the world price of oil fall below this level.

POSITION IN THE UNITED KINGDOM

Prior to October 1973 the United Kingdom energy policy aimed at running down gradually the domestic coal industry, and stimulating and encouraging the exploration and exploitation of the North Sea gas and oil resources. At the same time the degree of dependence on imported oil as a source of primary energy was increased, but a very heavy degree of protection was accorded to domestic energy producers, and above all to coal, by way of taxes which raised the home price of imported oil to a level just below that of domestically-produced energy.

This policy was not at all surprising. While the cost of imported oil (excluding excise duty) amounted in the late 1960s and early 1970s to between $2.30 and $3.00 per barrel, the cost of domestically-mined coal was around $3.50 per barrel of oil-equivalent, that of nuclear power (Dungeness B Power Station) to $3.60 per barrel of oil-equivalent and the estimated cost of North Sea oil was between $2.20 and $3.00 per barrel.

The effect of this policy was to produce, between 1962 and 1972, the production of domestically-mined coal from 197 (long) tons to 118 tons, to raise imports of oil and its products from 1.06m barrels per day (b/d) to 2.06m b/d, and to stimulate exploration and development of the North Sea deposits, with output from that source expected to reach between 1.7m b/d and 2.4m b/d by 1980.

New Energy Policy

Following the oil export restrictions by the Arab countries and the rise in the price of imported oil, the British government created a new Department of Energy (resurrecting the old Ministry of Fuel and Power), embarking implicitly on the creation of a new energy policy. Although the matter is complicated by the attempts on the part of the Community to develop a common energy policy, the general objectives of British policy can be identified without undue difficulty.

The bases of the new policy are the present and possible future prices of the main sources of energy, and the size of the likely indigenous resources in relation to the estimated expansion of demand. As regards the latter, the rate of growth of energy demand in the United Kingdom during the last twenty years or so approximated that of the GNP.

There is no doubt that, as in the United States, one of the aims of the new energy policy will be to reduce this correspondence by using the pricing mechanism, special fiscal measures and an educational campaign. It is very difficult to assess the effects of all such measures but it is quite possible that the energy/GNP growth ratio will in fact decline to 0.8/0.9, thus reducing both the rate of growth of energy demand as well as the absolute quantities that will be consumed during the second half of the 1970s and in the 1980s.

The present and possible future prices of the main sources of energy, shown in Table 7.3, indicate that in early 1975 all types of domestically-produced energy were cheaper than imported oil and that except for coal they could be expected to remain cheaper than imported oil and that except for coal they could be expected to remain cheaper in the second half of the 70s, even if it were assumed that the price of imported oil would decline to around $8.00 per barrel (c.i.f.).

TABLE 7.3

Costs of Main Sources of Energy (US $m per barrel of oil-equivalent)

	Spring 1975	*Say, 1978*[a]
Imported oil	about 12	8
Domestically-produced coal	about 10[b]	10
North Sea Oil	–	4–7
Nuclear	about 4	4

[a] In 1974 prices.
[b] Rough estimates after the miners' February 1975 award.

Looking at the possible demand for energy in 1980 this can be expected to increase by around 25 per cent over present levels, or by the equivalent of some 1.2m b/d of oil. If, as is increasingly accepted, the production of oil and gas from the North Sea could yield by 1980 around 2m b/d, thus virtually replacing existing imports, the equivalent of an additional 1.2m b/d of oil would have to be generated domestically in order to achieve energy self-sufficiency. One would have to rely on coal and nuclear energy for this additional requirement; the actual amount would depend on the acceptable level of energy imports.

The new energy policy appears to accept that some of the energy requirements will have to be met from imports. This is so because of the difficulties, technical and otherwise, of building up rapidly the nuclear power potential, the difficulties of expanding domestic coal capacity, and the ready availability of coal (from Poland) and oil (from Norway and other countries).

In broad terms it would appear that even with the crash domestic programme, the United Kingdom may have to import between 0.6m b/d and 1m b/d of oil-equivalent by 1980, thus reducing its dependence on imported energy supplies to between 15 and 25 per cent, unless the North Sea oil output exceeds 2m b/d. There are good reasons for thinking that not only might this be a conservative estimate but also that the growth in demand might prove less vigorous than assumed. Thus the gap to be filled by imported fuels could be further reduced, if not eliminated.

Implications of the New Policy

As in the case of the United States, the consequences of the new energy policy are considered under three main headings: behaviour of total output, its main components, and external account.

As regards the likely consequences of the new energy policy for the growth of total output and its main components, such estimates as can be made suggest that the total of additional capital spending that can be expected to be undertaken for energy expansion during the remaining years of the 1970s, assuming there to be a "crash programme", is likely to amount to between £29,000m and £35,000m (in 1974 prices). This comprises between £17,500m and £20,500m by other fuel and power industries as well as on additional infrastructure including transport.

The total of the prospective, additional fixed capital spending of between £29,000m and £35,000m (equivalent to between £4,000m and £5,200m per annum) compares with the annual average of gross capital expenditure for productive purposes (that is excluding dwellings and such like) over the last few years of some £8,000m per annum and implies a rise of between 55 and 70 per cent.

If it is also accepted that Britain's long-term growth potential is between 3 and 4 per cent per annum, equivalent to £2,150–£2,300m of additional output (at 1974 prices), the inescapable conclusion is that the new energy policy – in the absence of deliberate policy of financing a significant part of the new requirements from overseas – will absorb all of the additional production, leaving nothing for raising personal consumption and other types of spending, such as current government expenditure. Even if an appreciable part of the additional capital spending for energy purposes is financed from overseas, the United Kingdom can be expected to experience probably one of the most intensive and prolonged upswings of its history, which is likely to be associated with strong pressure on prices on the one side, and pressure on interest rates on the other.

The exploitation of the North Sea can be expected to bring very substantial benefits to the British balance of payments. The (gross) deficit on oil account (that is, not allowing for invisible earnings) in 1974 was some £4,500m following the rise in the world prices of oil. The United Kingdom may well break even on oil account by 1980 (on a gross basis) or at worst have a deficit (assuming the world price to be then around $8.00 per barrel) of some £700m – a substantial improvement over 1974.

POSITION IN THE ORIGINAL SIX

As mentioned before, the original six members of the European Community are very dependent, as far as primary energy is concerned, on imported oil and gas (see Table 7.1), which account for over 70 per cent of total energy resources. This large degree of dependence on imported energy is the result of deliberate policies pursued by the original six. These arose from the very high relative cost of indigenous fossil fuels, relatively small potential resources of oil and gas, whose cost in any case tended until recently to be higher than that of imported supplies, and of equally unfavourable costs of nuclear energy as compared with that of imported oil.

It has been estimated that the cheapest grade of coal in the "old" European Coal and Steel Community in 1967 was, at the mine, $16 per metric ton, to which there was added a subsidy of over $5, making the real cost $21 per metric ton. In oil-equivalent terms this represents some $4.50 per barrel as compared with prices of imported oil (c.i.f. Channel ports) of $0.3–$3.00 per barrel.

In these circumstances it was not surprising that each of the original six pursued its own energy policy, which in essence aimed at reducing the dependence on domestically-produced coal, raising the relative importance of imported oil, while at the same time placing a greater or smaller degree of emphasis on the introduction of nuclear power.

New Energy Policy

While the Brussels Commission has, for some years, tried to evolve a common energy policy, these efforts so far have been almost totally unsuccessful and each country before and after October 1973 has tended to follow its own policies. In view of the very small reserves of indigenous fossil fuels other than coal and limited potential for the development of oil and gas, the only alternative open to the members of the "old" Community, be it individually or jointly, was to try to expand the production of nuclear energy and wherever possible to exploit indigenous coal resources, to be used either in a traditional form or for gasification.

It should be stressed that because of the high cost of coal, which could be estimated at around $40 per ton in January 1975, or $9 per barrel of oil-equivalent, and further prospective rises in costs, the expansion of coal as a primary source of energy still involves a high cost. While at present, the price of imported oil amounts to $11–$12 per barrel it would seem, as argued before, that it could decline to around $8, a figure below the cost of domestically-produced coal. In these circumstances, further expansion of coal would represent primarily an insurance policy against subsequent increases in prices of imported oil, which may or may not happen.

To that extent the energy policy of the original six can be expected to concentrate primarily on the development of nuclear power and ensuring, by way of bilateral and other arrangements with the Arab countries, that the minimum quantities of imported oil are received at a reasonable price.

Implications of the New Policy

Such rough guesses as can be made about the additional capital spending to be undertaken by the members of the "old" EEC would indicate that it can be expected to amount to some $350,000–$450,000m and is likely to be concentrated predominantly on nuclear energy and to a certain extent on coal. This figure compares with the total of productive fixed capital spending (that is, excluding dwellings) of $120,000m per annum during recent years, implying a 30–50 per cent annual rise in outlays on productive capital formation.

During 1974 the net cost of imported oil to the original six helped to reduce the rate of demand for oil products (as the result of using the pricing mechanism and other measures) from around 8.5 per cent per annum experienced over the last ten years to some 6 per cent and, by 1980, might also reduce the degree of dependence on imported oil from around 64 to about 50 per cent, or indeed even to 40 per cent (this last figure is however highly optimistic). Even under these circumstances they would have to import approximately the same quantity of oil as in 1974 (this is, about 12m b/d). Even if the world price of oil declines by then to some $8 per barrel the gross cost of oil imports seems likely to be over $30,000m (at 1974 prices) as compared with around $6,300m in the twelve months ending in September 1973.

As a result, the original six will have to bear a large additional burden on their external accounts, which will require them to increase appreciably their exports to countries outside the Six.

It is this large and continuing dependence on imported oil which explains why countries like France are endeavouring to evolve long-term policy towards the Arab oil-exporting countries and place the trade relationship with them in the framework of bilateral "barter arrangements" which can be expected to play an increasingly important role.

POSITION IN JAPAN

Because of her small indigenous resources of fossil fuels, Japan's energy policy, like that of the continental European countries, has been to rely principally on oil as the main source of energy. As can be seen in Table 7.1, 75 per cent of Japan's primary energy is provided by oil, 99.8 per cent of which has been imported.

The remainder came from the domestic coal industry, the capacity of which was being run down, and from hydro and nuclear sources. The result of these policies was to raise the Japanese imports of oil in 1974 to nearly 5.5m b/d but, because of the special fiscal policies, this still kept the cost of energy as a percentage of total GNP at slightly above the United States level of 3 per cent.

New Energy Policy

Apart from arrangements designed to reduce the demand for

energy, by early 1975 the Japanese authorities had not revealed anything about the long-term energy policies they would pursue. Absence of indigenous resources, however, considerably restricts their freedom of choice. A modest expansion of the domestic coal industry and the exploitation of offshore oil and gas – where there are formidable legal and political problems – can be expected to cater for some of the existing and future requirements. Also, like other industrial countries, Japan is likely to embark on a very ambitious programme of expanding nuclear energy. This however cannot be expected to make a very substantial contribution in relation to total demand until well into the 1980s. Thus during the remaining years of the present decade the bulk of Japan's energy requirements will have to be obtained by way of importing oil, gas and coal.

Therefore almost inevitably the aim of Japan's energy policy will be to ensure that adequate supplies of oil and coal are obtained overseas. To this end Japan is likely to adopt a policy of barter and/or similar arrangements with the main suppliers of coal, oil and gas during the next ten years or so. Apart from the OPEC countries, with special attention being paid to Indonesia, Japan is likely to increase her imports of coal from Australia. Two other countries with whom increasing efforts are going to be made to reach agreement, will probably be China and the Soviet Union. In exchange for technical know-how and capital goods, Japan could probably obtain certain guaranteed quantities of energy which would enable her to ensure a minimum rate of expansion and to effect structural changes in the economy.

Implications of the New Policy

No official or private estimates have so far been published about the likely level of investment that will be required through 1980 in order that nuclear sources of energy may be increased whilst domestic coal, gas and oil capacities are being expanded.

Such very rough guesses as can be made lead to the conclusion that the additional capital expenditure that will have to be incurred within the new energy policy is likely to amount over the next decade to 9,000,000–11,000,000m yen (at 1974 prices) per annum, tantamount to about 50–65 per cent of 1974 fixed capital expenditure, excluding housing.

Additional capital spending of this order of magnitude is likely to maintain the Japanese rate of growth at a high level, though not as high as that achieved in the fifteen years from 1959–1974 when it amounted to 9 per cent per annum (in real terms). A lower rate of growth with relatively higher investment spending is bound to cause the authorities to adopt policies restricting the rate of growth of private and public consumption, resulting in pressures on interest rates.

Despite the efforts to expand indigenous sources of energy, Japan will continue to be heavily dependent on imports of oil (and also coal) with the result that the Japanese external accounts will be exposed to strong pressure and cause the Japanese authorities to adopt policies designed to ensure and secure supplies of oil.

In the six years from 1968 to 1974 consumption of oil in Japan expanded at an annual rate of around 13 per cent, with imports of oil (both crude oil and oil products) reaching nearly 5.5m b/d in 1974 at a gross cost of some $19,000m. If it is assumed that the Japanese authorities will suceed in reducing the rate of growth of consumption of imported oil in the next seven years to 5–6 per cent per annum, imports of oil by 1980 would nevertheless amount to 7–7.5m b/d. Even if it is accepted that the world price of oil will be around $8 per barrel (in 1974 prices) the additional cost of oil imports will be between $4,900m and $6,400m.

Thus the change in the energy situation is likely to place Japan in a relatively worse position, as regards external accounts, than that of other industrial countries, and can be expected to cause the Japanese authorities to try to negotiate barter and similar arrangements with energy suppliers.

LONG-TERM IMPLICATIONS OF THE NEW ENERGY POLICIES

The new energy policies of the main industrial countries aim to reduce the degree of dependence on imported oil and to develop domestic sources of energy supply. In order to do so they will all invest substantial amounts in this area. This policy of expanding investment can be expected to generate a very strong and prolonged upswing, resulting in a very high rate of expansion. As the size of investment needed will be in excess of the likely rate of growth, severe restrictions on other forms of spending can be expected, above all personal consumption and public sector current and capital spending. Considerable pressure will be exerted on interest rates and the rate of increase in prices.

Because the aims of the new energy policies differ as between the principal industrial countries, reflecting the varied potential resources of indigenous supplies, the effects of the new energy policies on domestic investment, on the rate of growth, and on balance of payments will be different. These are summarised in Table 7.4, which indicates that because of their ability to exploit indigenous oil supplies, the United States' balance of payments is

TABLE 7.4
New Energy Policies and their Consequences (figures in $1000m)

	United States	*United Kingdom*	*"Old" EEC*	*Japan*[a]
Additional capital spending (1974–80) in 1974 prices	1,000–1,250 or 150–200 p.a.	70–84 or 10–12 p.a.	350–450 or 50–60 p.a.	230–280 or 32–40 p.a.
Average annual capital spending other than dwellings, 1972–75	150	19	120	53
Growth potential, 1974–76 (in 1974 prices)	55–65	5–7	23–27	30–35
Balance-of-payments cost of imported oil in 1974	25 (gross)	(11 gross)	35 (net)	19 (gross)
Estimated balance-of-payments cost of oil in 1980 (in 1974 dollars)	negligible	under 2	30	25

[a] The United Kingdom and Japanese figures have been converted from the Sterling and Yen data, given in the appropriate sections of this chapter, at the end of February 1975 rates of $2.41 per £ and 285 yen per $ respectively.

likely to benefit very substantially while the costs of imported oil can be expected to bear heavily on the Japanese external accounts and that of the original Six. Because of these developments it would seem that the relatively competitive position of the United Kingdom can be expected to improve very significantly in comparison with that of the "old" EEC countries and Japan and even the United States, affecting in turn the position of sterling in relation to other principal currencies.

It needs to be stressed that overmuch importance should not be placed on the absolute size of the figures shown in the table. Their significance results from their relationship, and although their respective orders of magnitude are accurate, the peculiar uncertainties that present themselves to the forecaster in mid-1975 are likely to prevent exact accuracy in the estimates. They are however the logical outcome of the assumptions on which this paper has been based.

The relatively large degree of dependence of the "old" EEC countries and Japan on imported oil is likely to continue to have an adverse effect on their position in comparison with the United Kingdom and the United States. It will influence, as it has already done, their foreign trade and investment policies in the direction of bilateralism and will cause them to emphasise nuclear energy. All the industrial and political implications of such measures may therefore be expected.

The increased investment requirements will also require a large increase in savings. In the increasingly interdependent and integrated world economy, such savings will be obtained from domestic and foreign sources including the oil-exporting countries, whose import-absorptive capability will fall short of the rise in their external earnings, as well as international markets, whose existence is due to various restrictions of long-term and short-term capital movements.

In as much as savings, monetary policies and controls on capital movements differ among the various countries, international flows of capital can be expected to vary and have important repercussions on the structure of exchange rates and the international monetary system.

In addition to having profound long-term implications for the economic and political developments of the Western countries and indeed the whole Western World, the new energy situation and the new energy policies will pose formidable problems during the transitional period. These are essentially connected with the problems arising from the way imports of oil are financed during the next 3–5 years, and the way the oil-exporting countries decide to hold the balance-of-payments surpluses they accumulate. These problems are discussed in depth in Chapters 3 and 8.

NOTES

1. This is a revised and amended version of a paper given at a meeting of the British-North American Committee in May 1974.
2. See the discussion of the British position in Chapter 3 above, pp. 36–8.
3. At a time when energy policies are in a state of flux it is difficult to conduct an analysis of the kind attempted in this chapter. In the United States, for instance, the policy adopted by the Federal Energy Agency in January 1974 has been subject to continuing debate between Congress and the Administration. For the purposes of the analysis it therefore seemed best to take the position that existed in mid-1974.

CHAPTER 8

Higher Oil Prices and the International Monetary System

HARRY G. JOHNSON

In considering the impact of higher oil prices on the international economic order, the main implications for the international commercial system are discussed in Chapter 9, the subject of this chapter being the international monetary system. Whereas the problems discussed in earlier chapters have exacerbated issues already facing the General Agreement on Tariffs and Trade (GATT), necessitating major reforms, the "oil crisis" has not forced changes in the world financial structure. Indeed, if the stresses and strains of inflation had not already forced flexibility on the international monetary system, the rise in oil prices would have wreaked considerable economic havoc in a world still adhering to fixed exchange rates.

CURRENT SITUATION IN HISTORICAL PERSPECTIVE

Before discussing inflation, it would help to put the current situation in historical perspective, by taking a series of candid camera shots of the world – from the standpoint of the Western alliance – at roughly ten-year intervals beginning from the end of the Second World War. With the end of the war there was a period when, as seems in retrospect to be increasingly clear, the main objective was to reconstruct the golden age of nineteenth-century liberalism, assisted by the lessons of two world wars and the Great Depression. Economically the aim was to re-establish a liberal trading world, on the basis of an international monetary system more soundly proofed than before against depression and balance-of-payments crises.

It should be noticed that, by contrast to the situation in the middle 1970s, the overwhelming fear was world deflation. Thus the overwhelming response, institutionally, was to protect the employment policies of governments against the contagion of other governments' deflationary policies by permitting currency devaluations, subject to safeguards against "beggar-thy-neighbour" (competitive or offensive) devaluations, while retaining the general benefits of a world money in the form of a regime of fixed exchange rates between currencies. Correspondingly liberal trading rules meant minimising interference with imports, as well as restraining "artificial" incentives to export.

One should also notice that policies relating to international money, international trade and international investment in development were regarded as integral elements in a four-part package, the fourth element of which was meant to be policy relating to domestic employment and stability. Only the first three though were embodied in new international institutions: the International Monetary Fund (IMF), the General Agreement on Tariffs and Trade and the International Bank for Reconstruction and Development (IBRD). But, and again of relevance to the middle 1970s, it was not sufficiently appreciated that institutions develop personalities of their own. In short it was not realised that the three institutions would grow apart, their spheres of responsibility becoming isolated from one another, the interdependencies of their actions being more and more disregarded.

Similarly in the field of security, the perspective looked back to the nineteenth century, concentrating on the containment of imperial powers with territorial ambitions. For governments held to the obsolete belief that land and natural resources were the basis of national power, and therefore that the purpose and motivation of war was to acquire additional access to deficient or absent resources. The corollary of this belief was that continental powers – and to a lesser extent smaller nations already well furnished with empires – were inherently peace-loving.

Moving on ten years to the middle 1950s, through the unexpectedly prolonged troubles of postwar reconstruction in Western Europe, there was the unexpectedly rapid breakup of the colonial empires of Britain, France, the Netherlands and Belgium[1] and the generally unforseen emergence of rivalry between the continental superpowers – the United States and the Soviet Union – as the major threat to international security. The new security problem, characterised as the Cold War, transformed potential imperialist enemies into valued military allies, cosseted with Marshall aid; and superpower rivalry, combined with the disintegration of the European empires into congeries of new nations (since collectively categorised as the Third World), led to competition for political support from the emergent nations in the councils of the United Nations and elsewhere in exchange for lavish "development aid". Economically this changed perspective was reflected in, first, the gradual displacement of gold by the dollar as the basis of the international monetary system; and the support of that system by a ready, but not excessive, expansion of the supply of dollars and the willingness of the United States to tolerate the self-interested exploitation of the system by the countries of Western Europe and Japan at the expense of American trade and investment interests. Secondly, it was reflected in the subordination of the goal of a non-discriminatory multilateral (or worldwide) trading system to the goal of a European union based on regional discrimination – policies regarded as abhorrent when pursued by Germany in the 1930s but actively encouraged when deemed necessary for the achievement of "European unity".[2] Third, the changed perspective was reflected in the wholesale exemptions from the rules of fair and orderly competition that were afforded, in the name of "economic development of the Third World", to the "poor" and "less developed" countries, on whose behalf was exploited to an excessive degree the "infant industry" argument for protection as an exception to the classical presumption in favour of free trade.

Moving on a further decade to the middle 1960s, there was the thawing out of the Cold War in a *détente* between the United States and the Soviet Union, but the emergence of a concern over China as a threat to peace – reflected in the "domino theory" and the need for decisive Western intervention in the Vietnam war – and the beginnings of a rift between the United States and its West European, Canadian and Japanese allies.

Economically, and prior to the realisation of the inflationary effects of the escalation of the war in Vietnam, the resentment that existed over "dollar domination" that was inherent in the world dollar standard, as the IMF system had become, gradually settled into a realisation that the deficit of the United States was a special case. The view took hold that in the long run the way to get away from the dollar standard was to develop a new credit-based international monetary asset, namely Special Drawing Rights (SDRs) on the IMF. Regarding developments in Western Europe, the potentially divisive effects on the world economy of the European Community's formation, and especially its prospective enlargement to include the United Kingdom, came to be appreciated on both sides of the Atlantic. An attempt was made to check these effects in the Kennedy Round of GATT negotiations.

At the same time the *détente* between the United States and the Soviet Union had removed the purpose of their competing in the development assistance they offered the Third World, intercontinental weaponry having also removed to a considerable extent the purpose for maintaining military bases around the world. The emphasis of thinking on development policy had in any case begun to shift towards the extension of export opportunities and investment in new industries.

In this connection it is necessary to note two aspects of thinking in the Third World on this subject that had become significant by the mid-1970s. One was that to a significant extent the emphasis on the expansion of export opportunities, especially through tariff preferences in favour of developing countries, was simply an amendment to the previous emphasis on the "infant industry" argument to correct for the hopelessly inadequate size of the domestic markets as a basis for industrial development under heavy protection. The other aspect, involving the same mythical belief in the superior economic power of poor countries over rich ones, was that the argument for a "new trade policy for development" should include the exploitation of collective monopoly power in the production by developing countries of foodstuffs and industrial raw materials.

SECURITY ASPECTS OF THE OIL CRISIS

In coming to the present situation, and the new perspectives it demands, there are some security aspects of the oil crisis that might be disposed of first, if only sketchily and in amateurish fashion (since the present writer has no professional competence in this field).

It has been customary in postwar security thinking to distinguish between "nuclear" and "conventional weapons" warfare, and also to distinguish between "police action" or "extinguishing bush fires" and all-out warfare, with most of the attention going to the tactical study of means of confining police operations to the trouble spots, and fire-fighting to the bush, and to the strategic study of avoiding major warfare. But while the Middle East has long been recognised as a potential trouble spot,[3] it seems fair to say that almost everyone was caught by surprise by the resort to the monopolistic engineering of vastly increased oil prices, too, and to an artificial world oil shortage as a weapon of paramilitary activity.[4]

The real danger in that situation is closely parellel to what used to be thought – it turned out wrongly – to be the danger inherent in monopoly possession by the United States of the atomic bomb; the possession of invincible power by an authority used only to wield limited power within a sub-area of a system of checks and balances, and for purposes of defending and asserting its own self-interests against the comparably armed self-interests of other authorities. As it transpired in that case, at least in my judgement, both the United States (and the Soviet Union in is turn) have proved far more capable than was expected of assuming world responsibility in the sane exercise of power than was likely to have been the case had the power fallen into the control of any of the previous great powers whose modern successors made such a habit for a while of referring contemptuously to the "barbarism" of the two continental superpowers. Now there has developed a far more serious danger that small, feudally-governed, non-industrial nations suddenly able, through the caprice of the subterranean distribution of natural resources vital to the easy continuance of affluent industrial standards of living, to buy ultra-modern military armament in quantities sufficient to start a local war of ultra-bush-fire dimensions, will use this fortuitous change of circumstance to initiate a war for essentially feudal objectives that will draw the major industrial powers into conflict.

The world political system has been forcibly thrown back into nineteenth-century and earlier concepts of identifying the state with its territorial domain and citizenship with pre-emptive rights over landed property (in place of later twentieth-century concepts of the state as a social and economic community and enterprise and of skill and talent rather than citizenship or ethnic membership by itself as the basis of a claim for economic reward). The force of this reversal of political conceptualisation derives from the power of oil revenues and control of oil supplies to purchase alliance and vastly magnified leverage from myopic democratic governmental systems whose publics and politicians are prepared to mortgage the future heavily in order to avoid relatively small and short-run inconvenience to habitual standards of living.[5]

In these circumstances, it is easy – and, equally, futile – to urge the industrial countries to stand firm, cooperate, and not be "pushed around" by oil blackmail: "He encourages criminal behaviour by paying blackmail. You avoid embarrassment by paying hush money. I gladly make my fair contribution to the trivial expense of maintaining peace." What economic history has to teach about the general possibility of using cooperation in establishing rules of international economic relationship that will minimise risks to peace and security is both fairly clear-cut and extremely discouraging.

As a general proposition about the extreme possible test case – all the way back to Troy – siege, embargo and blockade have been outstandingly poor methods of achieving military victory, depending almost invariably either on catching the enemy off his home base in unfamiliar territory or on inducing treachery within his own ranks. At the level short of actual military conflict, and referring to contemporary history, the failure of the League of Nations to agree on oil sanctions against Italy in the 1930s, the failure of American-led efforts since the Second World War to coordinate and effectively enforce embargoes on trading certain types of goods with Communist countries (trade with Cuba being a conspicuous case) and the failure of Britain to achieve anything with her embargo on trade with Rhodesia: all three illustrate, in different ways, the difficulties of using trade as a political weapon, even when the value of the trade to the supplying country is not great.

The oil situation has two especially unpromising distinguishing features: (i) industrial countries are under immense pressure to increase their exports to the oil-producing countries by any means they can; and (ii) military equipment is in all likelihood the most profitable – with the highest price to cost ratio – kind of export available, owing to the high overhead cost of research and development and the low and generally falling marginal cost of additional production.

In view of these considerations the prospects of using international cooperation as a means of significantly reducing the security risks created by the sudden accession to enormous wealth by the oil-exporting countries, can only be regarded as dim. There are only two possible promising general avenues of action, both of them risky and difficult to achieve by cooperation. The shorter range one is to generate sufficient mutual poltical distrust among the oil-producing countries for them to abstain individually from policies endangering security in the Middle East and globally, as a condition of continued cooperation among themselves in exploiting the possibilities of oil monopoly. The second and longer range one is to erode the power of oil monopolisation by investing sufficiently heavily in conserving on the use of energy and finding alternative sources of energy. The endemic unwillingness of the democratic political process to face hard economic choices, and the consequent tendency to use anti-American sentiment to justify leaving the economic burden of such a policy largely to be borne by the United States, are likely to make the success of this policy dangerously dependent on the willingness and determination of the United States to bear more than its fair share of the economic burden of collective responsibility in return for less than its fair share of the benefits.

FIXED EXCHANGE RATES AND WORLD INFLATION

Turning to the monetary aspects of the international economic situation in the mid-1970s, it is necessary to understand the general history and nature of inflation.

Inflation has been the fundamental economic issue besetting the world for several years. It is important therefore to see the issue in perspective, and appreciate its linkage with the exchange-rate system and the international monetary system.[6] Until February–March 1973 (with the exception of a few brief "crisis" months in the second half of 1971) the postwar world was on a system of fixed exchange rates. Admittedly the exchange rates could be – and sometimes were – changed, but this was done with reluctance and only in cases of "fundamental disequilibrium".

Throughout this period barriers to the integration of the world economy (in the form of exchange controls and import controls) were being dismantled, or at least reduced and regularised, while improved communications techniques were steadily drawing the world as a whole into the domain of political and economic knowledge and decision-making. In short, economically the world was increasingly becoming one integrated market for goods, capital and labour. Evidence of this trend can be seen particularly in direct investment, which is being channelled through multi-national enterprises, and in the rapid development of a world market for liquid capital in the form of the Eurodollar market.

The integration of the world markets for goods and capital under a system of fixed exchange rates causes the world economy to behave like a single monetary economy. This means that if there is either deflationary or inflationary pressure exerted on a substantial scale at any point in the market, the result will be general deflation or inflation, the individual member of the system being unable to escape the general trend except by deliberate use of a floating exchange rate (or of repeated devaluations or revaluations), which would insulate its domestic monetary policy and price level from the world trend.

This phenomenon was appreciated only slowly and imperfectly, largely because governmental and policy thinking in the postwar period was shaped by wartime experience and specifically by the concept of a nation as an independent economic entity under the control of its own national government. A further reason is that the great crash of 1929 and the ensuing depression had strengthened belief in both the necessity and the power of independent national economic policy.

By the 1920s it had been recognised that inflation and deflation were worldwide phenomena, but the historical examples were of very long-drawn-out, and relatively mild, inflationary or deflationary epochs. The drastically quick and sharp international monetary deflation of the 1930s, made possible by the extent to which scarce gold has been eked out by fair-weather credit, was scarcely understood; and the world economy embarked on the post-Second World War era in the belief that the possibilities of world deflation had been removed by the alterations to the interwar gold standard incorporated in the IMF charter, and with little or no concern about world inflation as the converse to the problem of world deflation in the 1930s.

For some twenty years this confidence that the problems had been solved proved justified: the immediate postwar inflation, and the brief Korean War commodity price boom, could easily be explained as due to special and historically unique factors, while inflations in particular countries could be quite reasonably explained by national peculiarities that did not affect the world system as a whole. But underneath this favourable experience lay the fact that, by fortunate accident as opposed to deliberate economic management, the United States preserved a rough degree of domestic price stability. Given the size of the United States and its domination of the world system, especially having regard to the role of the American dollar as reserve currency for the other major currency countries, this meant world price stability, with other countries' price levels following the American price level, though with some room for marginal difference that could be accommodated by relatively small and infrequent exchange-rate changes.

All this changed with the American decision to escalate the war in Vietnam and to finance that war by monetary expansion instead of by increased taxes. Given the reserve currency role of the dollar, and the attachment of the other industrial countries to exchange rates fixed on the dollar, the inevitable result was world inflation. The other countries, to the extent that they recognised the nature of the problem, tended to waste their effort in lecturing the United States on the evil of its ways, complaining about the undue power to run balance-of-payments deficits that the reserve currency role gave the dollar, and propounding a scheme to remove this power by replacing the dollar standard with a system based on international credit money in the form of appropriately modified SDRs – instead of developing policies of using exchange-rate variation to insulate their economies from the American inflation.

Most of their attention however was devoted to efforts to restrain the national impact of world inflation by national monetary and income policies that were inevitably doomed to failure. On the more academic side, efforts were devoted to defending the idea that inflation was a national and not a world problem by showing, to their own satisfaction, that if international forces could affect an inflation presumed to be national in origin (by routes suggested by the national-inflation theory) the presumably relevant evidence did not support the hypothesis of an international inflation.

FLOATING RATES AND NATIONAL INFLATION

The important point with reference to the contemporary problem, is that the phenomenon of world inflation was crucially dependent on the fixed exchange-rate system, and that the hypothesis could no longer be maintained (at least in its direct form) once the fixed-rate system gave way to a regime of floating rates. Once exchange rates are free to float, national policy-makers regain control of their national money supplies and their rates of inflation in terms of domestic currency. In short there is no world money, or fixed exchange-rate system equivalent thereof, and hence no world phenomenon of monetary inflation.

The most one can do to justify the above hypothesis is to argue in one of two directions.

> It could be argued that the resort to floating rates occurred so long after the inception of the world inflationary process that expectations of inflation had become tightly geared to an expected continuation of world inflation; indeed that those expectations had become so tightly geared that national policy-makers were unprepared to embark on policies aimed at breaking expectations of national inflation and replacing them by expectations of stable national-currency prices and an upward floating exchange rate.
>
> Otherwise it has to be said that, in spite of the fact of floating rates, national policy-makers continued to hanker after, and think in terms of, a return to a fixed exchange-rate system. Thus in consequence either of historical views on the appropriate level of the fixed rate or of strategic thinking on the level at which the rate should be fixed in a restored fixed-rate system, they so behaved that the floating-rate system has been operated in reality as a fixed-rate system, albeit with a somewhat greater degree of latitude concerning – and uncertainty about – the variation in actual market rates and parities in terms of other currencies.

Both of these factors – (i) unwillingness to fight inflation boldly with the tools made potentially effective by floating rates and (ii) explicit or implicit acceptance of an eventual return to fixed exchange rates – undoubtedly underlie current opinion on the nature of the inflation problem and particularly the belief that it is a world problem requiring a cooperative world solution. In addition it has proved politically convenient to treat the drastic increase in the world price of oil – and to a lesser extent the relative rise in foodstuff prices since mid-1973 – as factors aggravating the world inflation problem. Neither development, being concerned with the relative price of a sub-set of commodities, has in fact anything at all to do with world inflation. But the pretence that they do enables national governments to avoid facing the need for rational and hard policy choices concerning how best to adjust to the real-income-redistributionary consequences of these relative price changes.

This deliberate confusion happens to suit not only the general interest of national governments in the veiling of real issues in political rhetoric, but also the specific interests of West European governments in passing the maximum possible share of the burden of adjustment on to the United States. At the same time it promotes the interests of the IMF in rebuilding a fixed-rate system in which its officials will possess the powers of control over the dispensation of financial largesse, at the expense of the richer and more responsible countries. And it serves the interests of the development aid lobby, and the alliance of developed-country intellectuals and developing-country politicians who attempt to turn every international problem into an opportunity for levying taxes on the consumers of the advanced countries, to be spent by the governmental bureaucracies of the less developed countries.

The problems of national inflation and of higher oil prices are one thing. The re-establishment of a fixed exchange-rate system – one feature of which is envisaged as large, regular, concessionary resource transfers from advanced to less developed countries under the control of bureaucrats in the United Nations agencies – is something quite different.

To consider the problem of inflation first, the remedy for inflation on a national basis (assuming a floating-rate system) is simply the standard one of deflating the economy to the extent – and, more important, for the period of time – necessary to break the habit of raising wages and prices in the confidence that sufficient demand – or more exactly a sufficient increase in the quantity of money – will be available to legitimate the inflationary increases.

The floating-rate system however complicates the political problem of facing this issue by making it a problem for national governments which cannot be passed off by blaming either the inflationary policies of other governments or allegedly unavoidable "world inflationary trends". It may also aggravate the technical problem of anti-inflationary monetary management. This is because the national public may continue to form its expectation of inflation of domestic wages and prices by reference to inflationary trends in other countries, regardless (at least initially) of the direction of domestic monetary policy. Also the initial effect of domestic tight money in raising domestic interest rates, relative to foreign ones, is likely to lead to inward capital movements that appreciate the exchange rate, so diminishing employment in exporting and import-competing industries. This could lead the public to protest against exchange-rate appreciation and demand an exchange-rate policy that is in effect a policy of encouraging continued inflation.

(In the longer run a successful price stabilisation policy, and its consequences in establishing an appreciating trend of the exchange rate, would lead to a *fall* in domestic interest rates relative to foreign. The lower interest rate would be compensated by the appreciation of the foreign exchange value of the currency or by, what is the same thing, a smaller rate of loss of real purchasing power of money.)

COMPLICATIONS OF ANTI-INFLATION ENDEAVOURS

Apart from the obvious problems of making the transition from a *nominally* floating exchange-rate system, which in reality is a slightly looser form of fixed-rate system, to a *genuine* floating-rate system used self-consciously to pursue an independent national anti-inflation policy, the problem of reducing inflation is particularly complicated for a variety of fairly obvious reasons. There are four in particular that can be easily identified.

One is that the increases in the relative prices of oil specifically and of energy generally, and of foodstuffs, involve a redistribution of real income away from industrial workers and urban consumers. The need for accepting the redistribution can be disguised by letting inflation continue, whilst a cessation of inflation (or at least rapid inflation) will require recognition of the need for a relative reduction in the real incomes of the industrial worker and urban consumer groups. It seems a safe generalisation that inflation is easier to stop when real incomes are continually increasing than when real incomes for important sectors have to be reduced, temporarily, until increasing productivity has time to fill the gap. Note incidentally that the argument has been phrased in terms of groups, not nations. Most nations aside from the OPEC oil producers will be redistributing income towards foreign suppliers, so that the nation as a whole will suffer a loss of real income. Note also that the rise in the relative prices of energy and of foodstuffs, which is deflationary in terms of real incomes, appears as inflationary in terms of money prices and "the cost of living", a fact which adds to public and political confusion about the problem. This is only one example of a general confusion about inflation. Rises in money prices are by themselves deflationary, not inflationary; they are a way of eliminating excess demand for goods and excess money supply. But popular thinking invariably regards increased taxes, interest rates and so on, which raise prices or reduce demand, as themselves inflationary. Price increases are a cure for inflation; the inflationary process consists of continuous re-creation of inflationary conditions requiring further price increases to correct them.

A second reason for the difficulty in overcoming inflation under current conditions is that to check inflation it is necessary to break inflationary expectations; and such expectations become more deeply ingrained the longer the experience of inflation lasts and particularly the longer that expectations of accelerating inflation are borne out by experience. Moreover in the modern world with its global news coverage and rapid communication, the experience of which one knows best is not necessarily one's own; and given the journalist's concept of news – which stresses bad news anywhere to the exclusion of good news or no news elsewhere – inflationary expectations are likely to take an unconscionably long time to break, even with an extremely severe dose of protracted deflationary policy.

The third difficulty is that human memories and perspectives are short and human attention fickle. Public attention may concentrate closely on one particular problem at a particular point of time, and shift to another problem long before the first one is solved. This difficulty has been only too apparent in various countries in recent years, with the public, through its political opinion processes, expressing approval of strong anti-inflationary action but becoming equally strongly disapproving of the resulting unemployment, high interest rates and other discomforts long before the medicine has begun to cure the economic patient of the inflationary disease. This problem becomes increasingly important as the time-dimension of a process of stopping inflation by deflationary measures creeps up on, and overtakes, the "honeymoon period" of a newly-elected government, with the result that a government that persisted with anti-inflationary policies would virtually guarantee itself electoral defeat while bequeathing the benefits of greater price stability to its political opponents.

A fourth factor helping to complete the vicious circle of contemporary inflation is associated with the fact that stopping inflation is a matter of taking actions that will be convincing as demonstrations that the government is determined to break inflation, and hence successful in breaking inflationary expectations. In principle inflation could be stopped easily, and with little pain, if the government announced that it would do everything necessary to stop inflation, meant what it said *and was believed by the public*. But if the government means it, and the public does not believe it, the public is in for a painful dose of deflation. And conversely, if the public believe it or half-believe it, but the government does not really mean it, the public will suffer losses through errors of decision based on erroneous expectations. It will therefore be important for the public to form a view on the credibility of government statements of intention to stop inflation. This will include a judgement both of determination and of ability to command support for the policy over the time required for it to be effective. And for reasons already given it has become decreasingly likely that the public can be persuaded that government mean it, in an effective sense, when they declare their determination to stop inflation.

All of this relates to a national effort to bring inflation to a halt. An as yet unstated question is whether it is worthwhile trying to do so, given that the floating rate removes the main cogent practical objection to inflation, namely its effects on the balance of payments and the international reserve position, and leaves the case dependent on its alleged adverse internal consequences. As a broad generalisation the most serious of these alleged consequences arise, not from inflation itself, but from the persistence and the implementation in policy of the belief in the constancy of the value of money. Thus for example governments oppose indexation of various forms of regular income payments and tax liabilities, even though such indexation is made desirable by the otherwise abitrary effects of government-created inflation. At the same time public opinion expresses vociferous indignation at the inflation-induced rise in house rents, mortgage interest rates and food, electricity and urban transport charges, even though on an overall balancing of rising money incomes against rising costs of living the allegedly injured are beneficiaries rather than losers.

(For example those who own houses financed by mortgages benefit substantially from inflation.)

The most serious real problem of inflation, in the author's judgement, arises from the fact that important sectors of economic activity – for example school-teaching, nursing, public service work, the civil service – are entrusted to professional and semi-professional people who customarily do not bargain about pay rates but assume that their pay will be "fair" for the job and permit them to maintain an appropriate standard of living. Inflation forces such people either to struggle to make ends meet or to transform their view of their standing from professional independence into wage-employee status. How well a bourgeois society can survive this proletarianisation of the professional classes is an open and serious question.

The problem of inflation is both fundamentally different in nature and far more complex and difficult if, instead of being viewed as overcoming domestic inflation for domestic reasons, it is envisaged as a problem of reducing rampant inflations proceeding at different rates in different countries to a common low "non-inflationary" rate involving sufficient similarity of national price trends to permit a return to a system of fixed exchange rates among the major currencies. That objective involves the additional two-fold problem (i) of restoring a system that even before the great inflationary surge began was showing increasing signs of strain, and was kept going by a commitment to cooperate in that endeavour which has since been virtually destroyed by the collapse of the system; and (ii) of setting a price-trend objective for member countries of the system that is dictated, not by national interest, but by commitment to an international objective to which national policy interests are to be subordinated.

I do not see any prospect in the visible future of a return to a fixed-rate system. Such a return seems to be dependent on there being a passage of time during which:

(a) the United States returns to a condition of reasonable price stability (there being little hope of this as the swing from anti-inflationary to anti-deflationary policy in the last few months of 1974 presages a lengthy period of stop–go and oscillatory inflationary experience);

(b) the re-establishment of domestic policy control over domestic money supplies, and the pursuit of price stabilisation policies in conformity with American trends, with sufficient success to enable the other major industrial countries to contemplate resumption of some sort of fixed-rate relationship with the United States.

"International co-operation" would then be required only at the stage of fixing the rules for changes in fixed exchange rates in cases of disequilibrium.

CONFUSED THINKING ON THE "OIL CRISIS"

Finally there is the problem of suddenly vastly increased oil prices and, more generally and gradually, of a substantial upward movement of energy costs: In popular and official discussions of this matter, there exists exactly the same sort of confusion between a fixed-rate system and a floating-rate system as is characteristic of the inflation discussion. Had the world still been on the previous fixed-rate system in the autumn of 1973, and subsequently, there would undoubtedly have been an international monetary crisis and probably a continuing period of international monetary chaos. But the troubles would have been associated with two somewhat different problems: (i) uncertainty about the levels of particular exchange rates appropriate to maintaining balance-of-payments equilibrium with the higher price of oil imports relative to manufactured exports; and (ii) the propensity of governments to stave off the need for adjustments of relative prices and production and trade patterns to the higher oil price, by spending their international reserves and borrowing additional international reserve assets.

It is undoubtedly vital to understanding of the "oil problem" to appreciate that, even on a fixed exchange-rate system, there is no inherent necessity for a sharp increase in the price of imported oil to lead to balance-of-payments disequilibra, deficits and international financial crisis. The occurrence of such phenomena has to be deduced from the presumed policy responses of governments. To see this one need only consider the question whether an individual faced with a rise in the price of gasoline for his car, or of electricity for his home needs, must necessarily (i) borrow on distress terms or (ii) borrow in the specific form of running down his cash balance. Obviously there is no reason why he needs to run a balance-of-payments deficit (run down his cash balance) or resort to distress borrowing. He can instead sell assets in the market, reduce his rate of saving (asset accumulation), reduce his expenditure on other goods or work harder and earn more.

A society differs from an aggregation of individuals only in that the various alternative possibilities of adjustment will probably require adjustments of relative prices, and in particular downward movements of some relative prices which under fixed exchange rates would require reduction in money prices and wages, that will be resisted strongly, given the habits of thought of a monetary economy. Governments are likely to prefer to finance continuation of activity without adjustment, and to run a balance-of-payments deficit, rather than confront the need for forcing domestic adjustment. Such resistance instead can be, and has been, avoided under a floating-rate system, to the extent that the question is one of "money illusion" in the technical phrase – that is, willingness to suffer a loss of real income through money prices rising relative to money incomes, but not through money income actually falling – though as mentioned the floating-rate system has enabled governments to rely on domestic inflation to disguise from the public the fact that real incomes somewhere must be reduced.

Under a floating-rate system a country still has a choice between reacting to an oil-import-price-increases-induced reduction in real income by cutting current consumption or by borrowing, in one way or another, against future increases in production and correspondingly potential increases in future consumption. It is important to note, because the point is usually overlooked,

(a) that this is a choice which individuals are perfectly capable of making for themselves if they have to do so;

(b) that there is absolutely no reason to think that, because governments have obvious political reasons for preferring to sustain current consumption levels by borrowing at the expense of future – and politically unrepresented – generations, their decisions in this regard are wise from the point of view of the ongoing society; and

(c) that there is still less reason to believe that international cooperation should be mobilised, under some principle of moral obligation, to encourage governments to borrow against the future by making international loans available to them on concessionary terms, at the expense of the oil-exporters or the richer industrial countries.

If the exercise of monopoly power is shameful, that is no reason why the conscience money should be paid by the monopolists to the most improvident and spendthrift countries; and if, contrary to fact, the rise in the price of oil is treated as a natural disaster of some kind requiring cooperative sharing out of the loss, there is no obvious sense in measuring the losses by the amounts of imports that national policies decide should be financed by borrowing. There is also the consideration that the worst possible way of discouraging the use of monopoly power in international relations is to insist that those who are hurt most by the exercise of monopoly power have a legitimate grievance, not against the monopolist, but against those who are also hurt but can afford it better.

The same sort of logic applies to the issue of so-called "recycling". There are really two issues here.

One is the old-style problem of recycling, discussed since the early 1960s: that in a fixed-rate system doubts about the sustainability of a particular country's exchange rate can lead to private speculation for or against the country's currency causing reserve flows which in turn may force exchange-rate changes, and that such flows can be neutralised by lending the money back to the central bank of the country from which it came. The difficulty with that solution was that when private money saw good reason to distrust an exchange rate, public (central bank) authorities found it hard to generate confidence enough in their own superior judgement to speculate on a large scale against market opinion. In any case this is a problem of an inappropriate set of rates in a fixed-rate system; in a floating-rate system the private speculators can be left to speculate against each other.

The new-style problem of recycling involves the fact that the oil-exporters are likely to want to invest most of their monopoly profits rather than spend them currently, and that these investments will be concentrated in a very few national capital markets, coupled with the presumption that this is an unfair situation which requires cooperation to redistribute the investment money "fairly" among countries.

That presumption is gratuitous. A major function of organised private capital markets is that of financial intermediation; that is, of providing securities of types that savers want in return for money which is reinvested in types of securities that investors in enterprises wish to issue, all in such a fashion that capital is invested for the best return and hence efficiently allocated among competing uses. Within a nation financial intermediation permits apartment-dwellers saving up for vacations and other rainy days to finance mortgages for home-owners, and also permits sedate purchasers of insurance to provide equity money for the brewers of beer. Between nations financial intermediation permits Canadian municipalities to build schools and hospitals with money borrowed from American bondholders, and French firms to lease equipment for production with money supplied by Canadian financial organisations. There is no reason to think that money invested by Arab sheiks in London property will not wind up financing tea exports from Sri Lanka, once the financial system gets used to the availability of such investable funds on a regular basis. The idea that a great effort of intergovernmental cooperation is necessary to make the normal financial processes perform is seductive but superfluous; and it lends itself only too easily to providing free rides for certain countries at the expense of others, including the investing oil-exporters.

There are two technical aspects of the idea of recycling that are worth specific comment. In the first place the popular mind, and even the official one, tends to be captivated by the tremendous danger of oil-importing countries piling up vast debts for current oil consumption that they will not be able to repay. But in fact, under current and prospective conditions, such debts will actually involve a transfer of real resources, properly measured, from lenders to borrowers. The rate of interest on the special oil facilities is 7 per cent; with a rate of inflation of say, 12 per cent (probably on the low side), the borrower is being paid 5 per cent per annum to take the loan (7 per cent nominal interest rate minus 12 per cent reduction in the real value of money). With this kind of subsidy on borrowing, sympathy for the poor helpless borrower is – as so often happens in inflation – completely misplaced.

Secondly, if the world is ever to return to reasonable price stability, this will require control over the growth of money and, in a fixed exchange-rate system, some kind of international control. Such control is obviously inconsistent with the controlling authority's regarding one of its main tasks as arranging for and lavishly extending credit to finance countries in living beyond their economic means. In this connection one is virtually forced to conclude that the IMF officials have signally failed to exercise the leadership in world monetary thinking that one would expect of them, and have instead allowed themselves either to be panicked by international political crisis or tempted by the opportunity to acquire political favour and support by abandoning principles in order to gain popularity.

To say this is not to say that everyone's sense of justice will be satisfied and, still less, that everything will immediately work smoothly and easily. It is to warn though against the idea that a potential crisis situation necessarily requires extensive new international institutions and, still more pointedly, to warn against the assumption that a new situation should be met by reforming old institutions according to old ideas of what was wrong with them.

NOTES AND REFERENCES

1. But not of Portugal's empire which was not broken up until the 1970s.
2. This objective is discussed in terms of "high" policy on the part of the United States in Richard N. Cooper, "Trade Policy is Foreign Policy", *Foreign Policy*, Washington, April 1973.
3. For an early politico-strategic analysis of the Soviet Union's interest in the Middle East, see the paper by Lionel Gelber (published in 1968) in the Atlantic Trade Study Programme, run by the Trade Policy Research Centre, London. The paper was republished as "World Politics and Free Trade", in Harry G. Johnson, *New Trade Strategy for the World Economy* (London: Allen & Unwin, 1969).
4. It might be noted, however, that an argument of those who, at the turn of the decade, favoured a continuing British military presence in the Persian Gulf was to ensure that no other major power could be in a position to deny oil supplies to Western Europe. See Geoffrey Williams, *Natural Alliance for the West*, Atlantic Trade Study (London: Trade Policy Research Centre, 1969).
5. The distinction between borrowing for consumption and borrowing for investment is underscored by W. M. Corden in Chapter 2.
6. For a general discussion on inflation see Johnson, *Inflation and the Monetarist Controversy*, De Vries Lectures (Amsterdam: North Holland Publishers, 1971).

CHAPTER 9

Higher Oil Prices and the Reform of the International Trading System

GEORGE F. RAY

Public discussion of the implications of the rise in oil prices has mainly focused on the financial implications, with intergovernmental action being taken by the International Monetary Fund (IMF) and, as far as developed countries alone are concerned, by the Organisation for Economic Cooperation and Development (OECD). It is in the OECD framework that the International Energy Agency has been established to carry out a comprehensive programme of cooperation – both in the event of emergency and over the longer term – among sixteen oil-consuming countries belonging to the OECD.[1]

But the rise in oil prices implies structural changes, not only within oil-consuming countries, but also externally; and those changes have implications for the reform of the international trading system, which has been regulated under the General Agreement on Tariffs and Trade (GATT) since the end of the Second World War.

STRUCTURAL CHANGES

The basic reason for these structural alterations is the change in relative factor costs due to the suddenly raised oil prices. Any substitute for oil – such as coal – will be in high demand; relatively energy-intensive products and services will become more expensive and their consumption will be relatively reduced in favour of less energy-intensive, and therefore relatively cheaper, products and services. The implications of such changes could be extremely far-reaching. To give a few examples: plastic-based packaging materials may be at a disadvantage as compared with other packaging materials, unless general inflation raises the price of the latter; aluminium's price relation to other metals or building materials will become less favourable; insulation materials will be used more widely and building standards might change; energy-saving forms of transport – even including the bicycle – will receive an impetus; alternative technologies will become more widely applicable; equipment and appliances which reduce industrial and other energy consumption (heat exchangers, high-productivity boilers and so on) will be in short supply as they will now have a very short pay-off period.

These shifts will bring about structural changes to the pattern of production and demand within each consumer country as well as internationally, and will result in contraction and unemployment in certain industries, and new jobs as well as relative boom in others. Countries specialising in energy-saving products will gain additional exports markets, whilst others will have to accept new import burdens. Consequently oil-consuming countries will need to adopt very different fiscal and monetary policies, measures concerning the exchange rate and the steering of their economies in general.

The structural shifts are likely to affect primary commodities as well. These are the main export-earners of most developing countries and it seems likely that in the near future demand for them, especially for industrial materials, may either fall or only rise slowly because of the reduced rate of growth in the industrial areas. But if the high price of oil should reduce the competitiveness of synthetic materials – and there seems little reason why it should not – then within an overall reduced aggregate demand there could be an increased demand for natural materials, such as cotton, wool, jute, rubber and hides. This change in circumstances is likely to prevent any major decline in the prices of such materials.

A quite different situation exists for metals and minerals, where the market has some similarities to that for oil, with a limited number of producers who, even in more normal conditions, have sometimes been successful in controlling the market by adjusting their production programme. Metal-mining and smelting require large quantities of energy.[2] Any shortage of energy could reduce output, and in any event the higher energy prices would affect production costs.

Though an oil-shortage might be expected to cause a slump or a reduction in demand through deflation, and a fall in commodity prices of the post-Korean War type might be anticipated on general orthodox grounds, a piecemeal analysis reveals little hard evidence for this. With all the uncertainties underlying this statement – in an area where forecasting has always been notoriously hazardous – it at least lightens the otherwise gloomy outlook for the less-developed countries facing higher import costs, whose income chiefly stems from producing and exporting primary commodities.

REFORM OF GATT ARTICLES

In view of the worldwide tendency towards chronic inflation, the development of severe shortages in the supply of many raw materials and the threat of a general recession, it has not only been the size of the transfer problem that has caused concern. What has mattered more is how the problem is managed by the industrialised countries. For in a very real sense the "oil payments" crisis has exacerbated stresses and strains in the international system of trade and payments that, during the late 1960s, had become plain for all to see.

How the transfer problem and the recycling of petro-dollars should be handled have been the subjects of previous chapters (especially Chapters 4 and 8), which is to say they will not be pursued any further in this paper.

As if the monetary problems are not daunting enough, the problems facing the international trading system are daunting as well, if in a different way. This is not the place to examine in detail the weaknesses that have developed in the GATT system over the last decade or so.[3] But it is worth recalling the role it has played in the past because of the effort that will have to be made by governments to strengthen multilateral cooperation in the future.

Following the disorders of the 1930s and 1940s, characterised by protectionist and discriminatory trading arrangements, the principles and rules of the General Agreement did much to restore during the 1950s and 1960s a degree of order in the conduct of international business. Indeed the liberalisation of international trade and capital flows in those years contributed greatly to a five-fold increase in world commerce, which in the 1960s was expanding twice as fast as world production.

Confronted with the worldwide dimensions of inflation and commodity shortages, governments are gradually coming to terms, however reluctantly, with the rapid integration of the world economy. If beggar-my-neighbour policies were an inappropriate response to the Great Depression, how much more inappropriate could they be forty years later, given the growing interdependence of national economies. For economic interdependence, and the trend for governments to assume ever wider responsibilities in domestic affairs, has increased the possibility of economic measures in one country seriously disrupting economic conditions in other countries.

In the late 1960s it was becoming noticeable that many GATT principles were being more honoured in the breach than in the observance. In particular the principle of non-discrimination set out in Article 1, requiring most-favoured-nation (MFN) or equal treatment to be accorded to all signatory countries, was being gravely eroded by a proliferation of discriminatory trading arrangements.[4] This was made possible by abuse of Article 24 which, under specific conditions, provided for departures from the MFN principle where customs unions or free trade areas were being formed between two or more countries. On other matters many provisions in the GATT had become inadequate to deal with developments in international trade, suggesting that they needed to be substantially revised or reformed.

The oil payments crisis has highlighted three particular aspects of the GATT system where multilateral consultation and negotiation needs to be established on a sounder basis. One relates to the threat of import restrictions, another to the provisions for structural adjustments in industry and yet another to the implications, both economic and political, of governments resorting more and more to bilateral economic agreements. There are other aspects that might be considered. But it is the above three that will be briefly discussed here.

When faced with a new and mammoth burden on the balance of payments, the orthodox approach has always been to try to save imports and promote an export drive. The tools are domestic deflation, import restrictions and export incentives, which may involve devaluation. All these might work in the case of any one country. But they could not work, and would definitely be extremely dangerous, in the present situation when almost all industrial countries face the same adverse circumstances. The deflationary impact of the high oil prices on their own, especially if part of the excess commitment is to be paid for in real terms by goods and services, is likely to be quite marked. If this is then supplemented by additional and purposeful deflationary measures in several countries, or all of them, the consequence might indeed be a world recession of considerable magnitude.

In May 1974 the developed countries agreed, within the OECD framework, on a temporary standstill on trade measures that might otherwise be introduced to offset the effects on balances of payments of the increased price of petroleum. But the agreement was only for one year. Such an agreement really needs to be maintained until more enduring provisions can be made. The fact is though that, as in the cases of Italy and Denmark, the economic situations of countries can deteriorate so much that governments are obliged to take actions that impinge on the export interests of other countries.

Article 12 of the General Agreement authorises countries in balance-of-payments difficulty to introduce quantitative import restrictions. Most industrialised countries however have now dismantled the administrative machinery for introducing quotas, but there is in any case a preference – exhibited most conspicuously by the United Kingdom in 1964 and by the United States in 1971 – to apply across-the-board an import surcharge. Perhaps provision should be made in Article 12 for this type of measure to be used.

There is though a more important consideration. The article was drafted, naturally enough, to cope with individual or isolated cases of countries in balance-of-payments disequilibrium. With the oil payments crisis the GATT system will have to cope with a large number of countries in deficit more or less at the same time. Since a much greater degree of flexibility has been written into exchange-rate relationships, the balance-of-payments rationale for import restrictions has been substantially removed, but in the circumstances of the oil payments crisis, compounded by inflation and the threat of recession, it may not be deemed appropriate to take measures that apply right across an economy. Instead governments might consider it more appropriate to apply trade restrictions on a selective basis, in order to "stabilise" or "fine tune" a floating exchange rate. Widespread resort to such a course could have a very disturbing effect on international economic relations. How could anyone be sure that trade restrictions justified publicly on payments grounds were not in truth being introduced for protectionist purposes? Accordingly there needs to be incorporated in Article 12, or in a separate protocol, a clear set of criteria and conditions for temporarily excusing a country on financial grounds from fulfilling its international obligations.

Mention was made earlier in this chapter of the structural changes in industry that are likely to ensue from the rise in oil prices. These are most likely to occur in the sectors of industry in which the developing countries have a comparative advantage. The identity of these sectors has become increasingly apparent over the last decade or so. But the developing countries have been encountering a spreading range of non-tariff barriers to their trade in the products concerned.

In the developed countries the increased level of protection has usually been explained in terms of giving the import-competing industry time to adjust, either by specialising more or by increasing its productivity so that it can again cope with open competition. Such temporary measures however have for the most part become permanent.[5] This is because the innovations of the import-competing industries, designed to raise their productivity, are soon introduced by the exporters of the developing countries. The research-and-development efforts of the former then have to be stepped up and so the process continues. As a result of the protection afforded by developed countries to industries competing with developing countries, a large part of the scarce research-and-development resources of the industrialised countries has been directed towards the creation of capital-intensive, and hence also energy-intensive, technologies for the manufacture of consumer goods that can be produced equally as well, and at lower cost, by labour-intensive means.

A recent report by a group of European economists concluded that it would be energy-saving to eliminate tariffs and let the low opportunity-cost manpower in developing countries produce standard consumer manufactures by simpler technologies. "In this way," the report added, "the research-and-development resources [of the industrialised countries] could be freed from a hopeless rear-guard action against a virtually inexhaustible reservoir of labour available at very low rates of pay. Those resources could instead be employed on genuine innovation for which there is now a sharply increased need."[6]

For these reasons and as a means of overcoming the transfer problem, the oil payments crisis has increased the need for developing countries to be accorded greater access to the markets of the developed countries, which therefore have to adjust internally. In contemplating a further expansion in trade between developed and developing countries, attention has to be paid to the problems of "market disruption" that might be expected to result, requiring the introduction of temporary import restrictions. With the possibility therefore of increasing resort to the General Agreement's Article 19, which provides for protection against sharp increases in exports of particular products, steps have to be taken to ensure that such measures are temporary. The energy crisis has thus increased the need to reform Article 19 which, as inferred above, has been ineffective in preventing "emergency" protection from becoming permanent.

How should Article 19 be strengthened? For a start, emergency protection should be introduced for a definite period, perhaps no more than seven years. Secondly, it should be regressive according to a prearranged timetable. Thirdly, it should be accompanied by a complementary programme of adjustment assistance in order to ensure that the protection is reduced on schedule. And fourthly, the whole process from beginning to end should be subject to international surveillance.

Turning to the third aspect of the GATT system, there have been a number of factors which have contributed to the erosion of the principle of non-discrimination, among them the "association" and other preferential trade agreements of the European Community, the development of "voluntary" export restraints mostly negotiated on a bilateral basis; and with the emergence of commodity shortages a number of countries are thinking of using export controls as a counter in negotiating access to the markets of other countries.[7] Unless the principle of non-discrimination is reasserted, there could be an outbreak of discriminatory trading arrangements among small groups of countries relating, perhaps, to small groups of commodities. Such a development would destroy the multilateral trading system to which small and middle powers must look for their interests to be protected.

It is a curious feature of the present situation that those countries which appear to be yielding to the deceptive attractions of bilateralism are also those who have so often been the champions and spokesmen of multilateralism. Britain and France have already concluded minor or major agreements with selected oil producers. Germany, Italy and Japan seem bent on following a similar course. (It is ironic that such an approach is being seriously considered and applied only a few years after the arch-bilateralists, the Eastern Bloc countries, started to move their trading arrangements towards a multilateral basis.) Any such action must weaken the bargaining position of the consuming countries.

The multilateral trading system has been operating relatively well in the recent past; even if just a part of the international petroleum trade was diverted into bilateral channels, the present trading system would be greatly disturbed, to nobody's lasting advantage. It could lead to a very unequal sharing of the non-renewable resources of oil. It would obscure market positions and prices, which would be a definite disadvantage to the oil-producing countries, and disturb established trade flows. It could easily put smaller countries into weak positions, result in unnecessary and unduly great pressure on some currencies, and lead to situations which could have a serious impact on the international economic mechanism, which will be in jeopardy in any case. In view of the closely intertwined nature of national economies, the inherent depressing forces of bilateralism may provide a strong downward multiplier, eventually reducing demand for oil to a level below that desired by any of the parties involved.

After the conclusion of one of these agreements, the British–Iranian deal, which was a very modest affair compared with the French deal a few days later, the European Community's Commissioner for External Affairs, Sir Christopher Soames, summed up his views in an outspoken attack on member countries who "surrendered to the dangerous temptation to make separate deals" with Middle Eastern oil states. Those who try to break ranks, he said, and fend for themselves, could "trigger off an ugly competitive auction of oil against money, oil against independence, oil against arms. If we cannot unite on this issue and face it together in loyal solidarity, the loss will not only be economic but could very quickly become political as well."[8]

The pressure resulting from higher oil prices will place considerable strains not merely upon relations between producer and consumer countries themselves. There is no doubt, if a major calamity is to be prevented, that the world's politicians must show much greater qualities of statesmanship and considerably more self-restraint over the next few years than they have usually exhibited in the recent past. There are too many signs that extremely powerful internal pressures on individual governments may tempt them to handle the situation from a nationalistic standpoint. The dangers to the European Community of such individual action have already been recognised; it has apparently been forgotten that if bilateral deals should be pursued more widely, no European country can hope to be able to offer a better deal to the oil producers than can the United States and perhaps even Japan.

NOTES AND REFERENCES

1. The sixteen members of the OECD's International Energy Agency are Austria, Belgium, Canada, Denmark, Germany, Ireland, Italy, Japan, Luxemburg, The Netherlands, Spain, Sweden, Switzerland, Turkey, the United Kingdom and the United States. Australia and New Zealand have both expressed an interest in joining, but Norway, a future oil-exporter, opted out of the oil coordinating group which established the agency.
2. This is partly due to technical development. At the start of the twentieth century most smelters would not work with ores containing less than 10 per cent copper. By 1940, with more advanced methods (which required more energy), 1 per cent metal content in the ore was considered feasible. In the mid-1950s 0.7 per cent metal content was acceptable. The situation with respect to other metals is similar.
3. Gerard Curzon, "Crisis in the International Trading System", in Hugh Corbet and Robert Jackson (eds), *In Search of a New World Economic Order* (London: Croom Helm, for the Trade Policy Research Centre, 1974) pp. 33–47.
4. On this theme, see Corbet, "Global Challenge to Commercial Diplomacy", *Pacific Community*, Tokyo, October 1971.
5. For a discussion of the shortcomings of the GATT safeguard mechanism and the reform of Article 19, see Jan Tumlir, "Emergency Protection against Sharp Increase in Imports", in Corbet and Jackson (eds), *op. cit.*, pp. 26–84; and also David Robertson, *Fail-safe Systems for Trade Liberalisation*, Thames Essay no. 8 (London: Trade Policy Research Centre, 1975).
6. Sir Alec Cairncross *et al.*, *Economic Policy for the European Community: the Way Forward* (London: Macmillan, for the Institut für Weltwirtschaft an der Universität Kiel, 1974) pp. 189–90.
7. In this connection, see Corbet, "Division of the World Economy into Economic Spheres of Influence", *Pacific Community*, January 1974.
8. *Sunday Times*, London, 27 January 1974.

APPENDIX 1

International Energy Agency of the OECD

On 15 November 1974 the International Energy Agency (IEA) of the Organisation for Economic Cooperation and Development (OECD) was set up with the aim of carrying out a comprehensive programme of cooperation both in the event of emergency and over the longer term.[1]

The energy crisis has clearly shown the need for international cooperation in this field. The Agreement on an International Energy Programme was the outcome of a joint endeavour to further cooperation among OECD countries. It is an effort to promote closer relations and solidarity. The programme follows the lines fixed by the preparatory work of the Energy Coordinating Group set up after the Washington conference held in February 1974.

To meet the aims which the participating[2] countries have set themselves, cooperation on energy matters must meet at least two requirements:

(a) it must make it possible to have a "horizontal" view of energy questions in all their aspects, whether technological, scientific, commercial, monetary, financial or political; and

(b) it must incorporate operational arrangements capable of providing a framework for cooperation which can rapidly be put into motion on the basis of efficient decision-making machinery.

The participating countries wanted the agency appointed to carry out the programme within the OECD – an organisation to which they belong and in which they have been cooperating for a number of years.

Thus the IEA was established as an autonomous body within the framework of OECD in response to a complex need: to maintain the political impetus and ensure the efficiency of the work, yet be able to draw upon the knowledge and experience of an existing organisation. Its aim is not to take over the functions of existing international organisations, but to facilitate their work. The fact that an entirely new international organisation has not been set up reflects the desire of the participating countries not to build up a system that would lead to confrontation with any country or group of countries, but rather contribute to a new system of international cooperation.

The International Energy Programme responds to the varied interests of the different parties. Some participants were more interested in long-term cooperation so as to diminish their dependence on imported oil. Others were primarily interested in working out discussions with the oil companies so that the market would be more transparent. Thus it was important to have an across-the-board agreement. This balance is reflected in the voting rules since no country or group of countries has a veto right. That is the important political factor of the voting scheme.

Participation in the IEA is open to all OECD countries which wish to join and are able and willing to meet the Agreement's requirements. It is also open to the European Community. The Agreement states that this does not in any way interfere with implementation of the treaties establishing the Community.

Cooperation with Oil Producers

From the outset the participants have shown special interest in cooperation with oil-producing countries and with other oil-consuming countries. Their aims in this regard are confirmed in the Agreement and this cooperation is meant to be one of the IEA's main activities.

Having established this principle, the IEA must go further and establish a dialogue. The aim of the dialogue is to move from a unilateral system to one in which consultation and discussion permits results which work to the advantage of all interested parties. But if there are to be results, there must not be only producers and industrialised consumers around the table, but also consumers from developing countries.

Long-term Cooperation

One of the fundamental problems is to lessen the dependence of the participating countries on oil. This is not so much a question of the industrialised countries becoming independent as their having a better energy balance since a goal of total independence would mean that the countries participating in the Agency assume nothing can be achieved in the dialogue with the producing countries; and this, very evidently, is not the case.

The scope for group action on long-term cooperation is particularly vast, since in addition to research and development in ten priority areas, it includes rational use of energy, uranium enrichment and the development of alternative sources of energy, while not forgetting protection of the environment.

In a determined effort not to allow long-term problems to be neglected, the Agency sets a deadline by which decisions must be made on proposals set forth.

Sharing in an Emergency

In addition to these activities the intention of the participating countries was to take out a sort of insurance policy against supply difficulties which might occur. That is why they have established a programme of emergency measures, including a system for sharing oil in an emergency. This system is precise, strict and detailed, but the agency's administrators hasten to stress that it is by no means the *raison d'être* of the IEA. The sharing agreement is based on equal sacrifice for everybody, and every participant must make a comparable effort. This is worked out in great detail and is very highly organised from a technical point of view. The oil is pooled under the control of governments and thus the system is based on the political responsibility of governments.

Thus, in the event of an emergency, the scheme will attempt to solve the difficulties of participating countries but it is claimed that it will not fail to take into consideration the needs of other nations.

Cooperation with the Oil Companies

The provisions for cooperation with oil companies go beyond what is required in an emergency, since they aim to establish within the IEA a permanent framework for consultation, and an information system which will include a general section (i) on structures and activities of companies, and (ii) on costs and prices.

Meetings of representatives of the Coordinating Group and the major oil companies are said to have augured well for future initiatives in this direction. Collaboration with the companies responsible for supplying and distributing oil in each country means in effect having a dialogue with market forces. The oil companies must follow a policy which is coordinated with that of governments.

PROGRAMME FOR INTERNATIONAL ACTION

The countries participating in the International Energy Agency have drawn up an International Energy Programme. This programme encompasses: (i) an allocation scheme in times of emergency, including emergency reserve and demand restraint obligations; (ii) an extensive information system on the international oil market; (iii) consultation with oil companies; (iv) long-term cooperation on energy; and (v) relations with producer countries and with other consumer countries.

Emergency Self-sufficiency, Demand Restraint and Allocation

In the event of a cut – actual or expected – in oil supplies in any one or more of the 16 countries which form the IEA, the group as a whole will share the burden of the affected nation or nations.

According to the programme, each participating country shall maintain emergency oil reserves sufficient to sustain consumption for 60 days with no net oil imports (later to be increased to 90 days). In addition, each participating country must at all times have ready a programme of contingent oil demand-restraint measures which would be activated in an emergency and which would be sufficient to reduce consumption by the amounts specified in the Agreement.

The emergency allocation features of the Agreement can be summarised in a simplified fashion: Allocation of oil and mandatory demand-restraint measures will take effect when either (a) the group as a whole sustains, or can be reasonably expected to sustain, a 7 per cent reduction[3] in its oil supplies or (b) one (or several) countries sustains, or can be reasonably expected to sustain, a 7 per cent reduction in its oil supplies even though oil supplies to the group as a whole have not declined by 7 per cent.

In the latter case, the country affected must first reduce its own consumption by 7 per cent. The rest will be made up by the other members on the basis of their consumption and through means of their own choosing, including demand-restraint measures or use of emergency reserves.

In the former case, each country shall implement demand restraint measures sufficient to reduce oil consumption by 7 per cent (by 10 per cent if the shortfall of oil supplies to the group is 12 per cent or more)[4] and available oil will be allocated among the participating countries according to a predetermined formula based on each country's permissible level of consumption (after activation of emergency demand-restraint measures), emergency reserve drawdown obligations, actual net oil imports, and normal domestic production.

If the cut in supplies continues long enough or is severe enough to use up half the emergency reserves of the 16 countries, new measures (further demand-restraint is mentioned in the Agreement as one possibility) will be proposed by the governing board of the IEA.

Each country's plans and programmes for emergency reserves, demand-restraint and allocation will be reviewed and assessed by the IEA, which may suggest means to enhance the effectiveness of the national measures.

The whole emergency procedure will be carried out in cooperation with the oil companies, and the oil will, in so far as possible, be distributed through normal channels and at market prices.

It is not an objective of the programme to seek to increase, in an emergency, the share of world oil supply that the group had under normal conditions, nor does any aspect of the allocation programme preclude any participating country from maintaining exports of oil to non-participating countries.

Speed is the essence of these emergency measures and deadlines are established in the Agreement for each stage of the emergency response.

Information System on the International Oil Market

An extensive information system will be developed and operated on a permanent basis by the OECD Secretariat. It will have two parts: one covers the situation in the international oil market and activities of oil companies; the other is designed to ensure the efficient operation of the emergency measures mentioned above.

General information relating to oil companies to be collected covers: corporate structure; financial structure, including balance sheets, profit-and-loss accounts and taxes paid; capital investments realised; terms of arrangements for access to major sources of crude oil; current rates of production and anticipated changes; allocations of available crude supplies to affiliates and other customers (criteria and realisations); stocks; cost of crude oil and oil products; prices, including transfer prices to affiliates; and other subjects, as decided by the governing board.

The data connected with emergency measures to be collected

includes: oil consumption and supply, demand-restraint measures, levels of emergency reserves, availability and utilisation of transportation facilities, current and projected levels of international supply and demand, and other subjects, as decided by the governing board.

This information will be held confidential and will be presented in such a way as not to harm competitive interests. Its collection will require legal action in a number of countries. Excluded from the system will be information about patents, trademarks, scientific or manufacturing processes, individual sales, tax returns, customer lists, geological and geophysical information or maps.

Consultation with Oil Companies

The agency will act as a framework for consultations with the oil companies so that governments can have an overall picture of the oil market and eventually undertake cooperative action with them.

Long-term Cooperation on Energy

In order to reduce the dependence of participating countries on imported oil, the members of the agency will undertake cooperative programmes, particularly in four main areas:

(a) energy conservation;

(b) the development of alternative energy sources – domestic oil, coal, natural gas, nuclear energy and hydroelectric power;

(c) energy research and development including as a matter of priority cooperative programmes on coal technology, solar energy, radioactive waste management, controlled thermonuclear fusion, production of hydrogen from water, nuclear safety, utilisation of waste heat, conservation of energy, utilisation of municipal and industrial wastes so as to conserve energy, and overall energy systems analysis; and

(d) uranium enrichment.

Relations with Producer and Other Consumer Countries

The participating countries will endeavour to promote cooperative relations with oil-producing countries and with other oil-consuming countries, including developing countries. They will keep under review developments in the energy field with a view to identifying opportunities for and promoting a purposeful dialogue, as well as other forms of cooperation, with producer countries and with other consumer countries.

To achieve these objectives the participating countries will give full consideration to the needs and interests of other oil-consuming countries, particularly those of the developing countries.

The participating countries will seek the opportunities and means of encouraging stable international trade in oil and of promoting secure oil supplies on reasonable and equitable terms. Other types of cooperation may include accelerated industrialisation and socioeconomic development in the principle producing areas and other energy questions of mutual interest, such as conservation of energy, the development of alternative sources, and research and development.

NOTES AND REFERENCES

1. This note is based on an article in the *OECD Observer*, Paris, January–February 1975.
2. Austria, Belgium, Canada, Denmark, Germany, Ireland, Italy, Japan, Luxemburg, the Netherlands, Spain, Sweden, Switzerland, Turkey, the United Kingdom and the United States.
3. Calculated using average final consumption of oil during a base period which is defined as the most recent four quarters with a delay of one quarter, necessary to collect information.
4. A country may substitute for demand-restraint measures use of emergency reserves which are in excess of its emergency reserve commitments.

APPENDIX 2

Assessment of Long-term Energy Development and Related Issues

In May 1972 the Council of the Organisation for Economic Cooperation and Development (OECD), meeting at ministerial level, called for an assessment of long-term energy prospects. The resulting report was published in January 1975.[1] The study drew on contributions from specialised bodies of the OECD such as the Energy Committee, the Oil Committee, the Environment Committee, the Committee for Scientific and Technological Policy and the Nuclear Energy Agency of the OECD.

Projections of energy supply and demand for the OECD area were based on a range of assumed prices for imported crude oil in 1980 and 1985. (These assumptions ranged between $3 and $9 f.o.b. in constant 1972 United States dollars.)[2] They assumed no change in the policies of member governments in energy and related fields, other than those necessary to ensure the effective working of the market mechanism, and are linked to forecasts of GNP growth over the period 1972–85 made before the large oil price increase and subsequent economic slowdown.

The projections indicate that if there is no substantial change in the real prices of imported crude oil from their end-1974 level, then:

(a) the overall energy consumption would grow at an annual rate of 3.5–4 per cent up to 1985 compared with the 5 per cent expected prior to the large increases in the oil price;

(b) the OECD area would produce almost 80 per cent of its energy requirements by 1985 (this proportion having previously been expected to decline to 55 per cent, from 65 per cent in 1972);

(c) the share of oil in total energy consumption would fall from 55 per cent to around or below 45 per cent by 1985; and

(d) as a result of reduced growth in overall energy consumption and higher indigenous production of oil and other forms of energy, OECD oil imports would be considerably below present levels, particularly after 1980.

The broad economic and financial implications of the projections based on the $9 assumption are the following:

(a) The rapid expansion of high-cost domestic energy production, together with continued high import prices, would add significantly to the overall cost level in the OECD area;

(b) The higher cost of oil imports from OPEC countries will entail substantial financial transfers. Assumptions concerning the proportion of these transfers that will be actually spent on imports of goods and services are inevitably rather speculative. The OPEC current surplus could be of the order of $30,000m in 1980 with a cumulative surplus from 1972 to 1980 of $250–325,000m (in constant 1974 dollars);

(c) The transfer of real resources corresponding to these assumptions would be equivalent to 1½ per cent of the OECD's GDP projected for 1980, after a gradual increase over the next six years. This figure, measuring the shortfall in real income as related to real output, would vary considerably from one country to another, the figures being the highest for Japan and some countries in Western Europe, whereas Canada and a few other countries would actually benefit from the change in terms of trade.

The estimated total potential for energy conservation (without recourse to emergency demand restraint) in OECD countries is 15 to 20 per cent of consumption levels previously forecast for 1985. The effort required would be much greater for some countries than for others.

The potential for the development of OECD indigenous energy resources is high, even with presently available technology. Taking 1972 as a basis, it is thus estimated that by 1985 oil and gas production could be doubled, coal production increased by 65 per cent (and more if markets could be found), hydro- and geothermal power by 67 per cent, and nuclear power increased twenty-fold (to provide up to 34 per cent of electricity generated). The production potential is very unevenly distributed within the OECD area.

Achievement of the estimated potential for energy conservation and production would make the OECD almost self-sufficient in energy by 1985. But the capital investment required would represent a very large shift of resources into the energy sector, particularly for countries rich in fossil fuel resources. Realisation of this potential therefore may be in conflict with other economic objectives, and may consequently be undesirable.

Little has been said in the body of this book about energy conservation. Yet it clearly requires further consideration. The OECD defines energy conservation as "the reduction in the amount of energy consumed without significant reduction in gross domestic product, general standard of living or level of personal comfort". It can take the forms of wasting less energy, increasing efficiency of energy conversion and end use, and changing certain living patterns. An analysis of the main areas in which energy might be conserved appears in the following table.

NOTE AND REFERENCE

1. *Energy Prospects to 1985* (Paris: OECD Secretariat, 1975).
2. $9 corresponded to about $10.80 in end-1974 prices.

TABLE I
Energy Conservation Possibilities

Type of conservation	Potential savings in % of sectoral consumption (per year)	Costs–Benefits	Type of policy action available
I. ELECTRICITY CONVERSION			
Improve efficiency of generation and transmission			Encourage R & D by utilities
(a) Smoothing daily demand cycle and reduction of peak load	Slight increase in generation efficiency	Lower capital requirements for new plant: possibility of earlier retirement for old plant	Revised tariff structures (less favourable rates for heavy users)
(b) Waste heat utilisation (total energy systems and district heating)	10% increase in efficiency (1985) Total energy systems use 15% less primary energy than systems with fuel heating and electrical cooling	Less economies of scale for small plants – less thermal pollution	Removal of legal and institutional obstacles. Encouragement of initiatives by utility companies
II. INDUSTRY			
Eliminate waste:			
Less non-productive idling of plant	10% of fuel	Small cost compared to savings	Information and encouragement
Better regulation and control Heat recovery	5% electricity (immediately)		
Replacement of old equipment and processes by more efficient ones Improved thermal insulation Increased maintenance of energy-using equipment More heat recovery systems	5% (1985)		Ensure that energy price to industry reflects real long-run marginal costs of production Accelerate efficiency improvements by tax incentives and/or credit facilities

Recycling of selected materials	Scrap processing requires 5% for aluminium and copper and 15% for steel of energy required to process ore	Scrap collection system Recycle-oriented design of products Less environmental pollution	Subsidy for R & D taxes or lower depletion allowances on virgin material
III. TRANSPORTATION			
Driving at lower speeds	25% less fuel consumption at 80 km/hour than at 115 km/hour	Reduction in accidents involving injury or death	Speed limits
Switch to smaller cars Automobile efficiency improvements:			
Redesign of body	5% } Less specific fuel consumption	Small cost	
Change to radial tyres	10% }	Small cost/longer tread life, greater safety	Graduated tax on fuel economy
Improved load to engine matching	15–20% }	$100–$200 } Less engine wear	
Installation of overdrive	20% during use } Switch to smaller cars and efficiency improvements estimated to save 6% (1980) and 22% (1985) of transportation sector fuel	$150 }	
Increased commuter car pooling	1–2% (1985)	Reduces congestion	Encouragement of reduced road tolls for cars containing 2 or more occupants during commuting hours
Increased use of public transportation in cities:	Depends on present state of development of public transport system	Reduces traffic congestion and pollution	High city centre parking fees and fines

TABLE I (*continued*)

Type of conservation	*Potential savings in % of sectoral consumption (per year)*	*Costs–Benefits*	*Type of policy action available*
More priority bus lanes		Low cost: discourages private cars	Subsidies to public transport
Fast commuter bus services		Low cost: discourages private cars	
Extension of rail and subway systems		High investment	
More cycle paths		Low cost: increased safety Encourages switch to bicycle for short trips	
Introduction of fast intercity train service		High investment: competes with automobile and aeroplane for trips of less than 500 km	Subsidy
Switch some intercity freight to rail	1–3% (1985)	Investment in terminal handling facilities/competes with road haulage	Subsidy
Increased load factors in airplane flights	22% jet fuel savings (1980) } 2.5% of transportation sector		Encouragement of co-operation between airlines
Lower plane cruising speed	3% } 2.5% of transportation sector		
Longer Term Aspects			
New propulsion technology			Support for R & D
Development of urban clusters			
Advanced communications systems to save trips			

IV. RESIDENTIAL – COMMERCIAL			
Adjustment of thermostats by 1°–2°C	6% per degree of space heating–cooling requirements (immediately)		Encouragement
More energy-conscious construction	30–40% of space heating needs in new buildings or	2–3% of cost of new building	Revised insulation standards for new buildings: financial incentives for insulation of existing buildings
Improved thermal insulation and temperature control	20–25% of sectoral consumption (1985)		
Increased use of heat pumps	5–10% of space heating needs, depending on degree of insulation		
Heat recovery from ventilation stream			
More efficient appliances	1–2% of sectoral consumption (1980)	Higher initial cost/lower running costs	Efficiency labelling requirement, possibly tax on efficiency
Reduce excess lighting	4% of total electricity contion		Encouragement
District heating schemes (using waste heat from electricity generation or burning municipal waste)		Reduces thermal pollution	Removal of institutional barriers, grants for pilot schemes
Increase R & D on better design, and operation of buildings, and on appliance technology	Increases the probabitity of achieving maximum potential	Accelerates the impact of new energy-saving technology	Support for R & D

Source: *OECD Observer*, Paris, January–February 1975, pp. 6–7.

Selected Bibliography

The bibliography below has been prepared, for both the specialist and the general readers, in an attempt to provide a useful list of more up-to-date literature on various aspects of the oil situation. The works are grouped into three sections: general background, which includes sources of statistical information on oil; the economics and politics of the oil crisis, which in a book of this nature necessarily constitutes the largest part of the bibliography; and alternative sources of energy, a field which has since 1973 received considerable attention. This bibliography does not in any way represent an effort to provide a comprehensive list, for the vast amount of literature on this subject precludes such a possibility.

GENERAL BACKGROUND

Energy Statistics (Brussels: European Community Statistical Office) published annually.

J. E. Hartshorn, *Oil Companies and Governments* (London: Faber & Faber, 1967).

Gerald Manners, *The Geography of Energy*, 2nd ed. (London: Hutchinson, 1971).

Statistical Review of the World Oil Industry (London: British Petroleum Company) published annually.

Christopher Tugendhat and Adrian Hamilton, *Oil: the Biggest Business* (London: Eyre Methuen, 1975).

UK Digest of Energy Statistics (London: HM Stationery Office) published annually.

World Energy Supplies (New York: United Nations) published annually.

ECONOMICS AND POLITICS OF THE OIL INDUSTRY

M. A. Adelman, *The World Petroleum Market* (Baltimore: Johns Hopkins Press, for Resources for the Future, 1972).

W. A. C. ADIE, *Oil, Politics and Sea-power: the Indian Ocean Vortex* (New York: Crane, Russak, for the National Strategy Information Centre, 1975).

JAMES E. AKINS, "The Oil Crisis: this Time the Wolf is Here", *Foreign Affairs*, New York, April 1973.

LORD BALOGH, "British Policy in the North Sea", *Banker*, London, March 1974.

LORD BALOGH, "North Sea Oil: a Rebuttal", *Banker*, London, September 1974.

"Can the Euro-currency Market Finance the Oil Deficits?", *Banker*, London, November 1974.

Capital Investments of the World Petroleum Industry (New York: Chase Manhattan Bank) published annually.

GEOFFREY CHANDLER, "The Changing Shape of the Oil Industry", *Petroleum Review*, London, June 1974.

GEOFFREY CHANDLER, *Oil Prices and Profits* (London: Foundation for Business Responsibility, 1975).

PHILIP CONNELLY and ROBERT PERLMAN, *The Politics of Scarcity* (London: Oxford University Press, 1975).

Energy: from Surplus to Scarcity?, Proceedings of the Institute of Petroleum's summer 1973 meeting (London: Applied Science Publishers, for the Institute of Petroleum, 1974).

Energy R and D Problems and Perspectives (Paris: OECD Secretariat, 1975).

JOHN E. GRAY, *Energy Policy: Industry Perspectives* (Cambridge, Mass.: Ballinger Publishing, 1975).

J. P. HAYES *et al.*, *Terms of Trade Policy for Primary Commodities*, Commonwealth Economic Paper no. 4 (London: Commonwealth Secretariat, 1975).

ROBERT E. HUNTER, *The Energy "Crisis" and US Foreign Policy*, Development Paper no. 14 (Washington: Overseas Development Council, 1973).

The Increased Cost of Energy: Implications for UK Industry (London: HM Stationery Office, for the National Economic Development Office, 1974).

ABDUL AMIR Q. KUBBAH, *OPEC Past and Present* (Vienna: Petro-Economic Research Centre, 1974).

WALTER J. LEVY "World Oil: Cooperation or International Chaos?" *Foreign Affairs*, New York, July 1974.

ASHRAF LUTFI, *OPEC Oil* (Beirut: Middle East Research Centre, 1968).

RICHARD B. MARCKE, *Performance of the Federal Energy Office* (Washington: American Enterprise Institute, 1975).

RAYMOND F. MIKESELL *et al.*, *Foreign Investment in Petroleum and Mineral Industries* (Baltimore: Johns Hopkins Press, 1971).

ELIZABETH MONROE and ROBERT MABRO, *Oil Producers and Consumers: Conflict or Cooperation* (New York: American Universities Feld Staff, 1974).

PETER R. ODELL, *Oil and World Power: Background to the Oil Crisis*, 3rd ed. (Harmondsworth: Penguin, 1975).

The Oil Import Question: a Report on the Relationship of Oil Imports to National Security (Washington: US Government Printing Office, for the US Cabinet Task Force on Oil Import Control, 1970).

"Oil and International Debt", *Banker*, London, April 1974.

"Oil and Money", Oil Supplement to the *Banker*, London, March 1974.

EDITH T. PENROSE, *The Large International Firm in Developing Countries: the International Petroleum Industry* (London Allen & Unwin, 1968).

COLIN ROBINSON, *The Energy "Crisis" and British Coal*, Hobart Paper no. 59 (London: Institute of Economic Affairs, 1974).

ANTON B. SCHMATZ, *Energy: Today's Choices, Tomorrow's Opportunities* (Washington: World Future Society, 1974) with a foreword by Gerald Ford.

GEORGE W. STOCKING, *Middle East Oil: a Study in Political and Economic Controversy* (London: Allen Lane, The Penguin Press, 1971).

JACOB VINER, *International Economics* (Chicago: The Free Press, 1951).

JOSEPH YAGER and ELEANOR B. STERNBERG (eds), *Energy and US Foreign Policy* (Cambridge, Mass.: Ballinger Publishing, 1974).

STANISLAV M. YASSUKOVICH, *Oil and Money Flows: the Problems of Recycling* (London: Banker Research Unit, 1975).

Coal and Energy Quarterly, London, Summer 1974.

ALTERNATIVE SOURCES OF ENERGY

GEOFFREY CHANDLER, "Energy: the International Compulsions", *Energy, Europe and the 1980s*, International Conference of the Institution of Electrical Engineers *and others*, May 1974.

"Energy and Power", *Scientific American*, special issue, New York, September 1971.

Energy Prospects to 1985: an Assessment of Long-term Energy Developments and Related Policies (Paris: OECD Secretariat, 1974).

PETER HILL and ROGER VIELVOYE, *Energy in Crisis: a Guide to World Oil Supply and Demand and Alternative Resources* (London: Robert Yeatman, 1974).

H. C. HOTTEL and J. B. HOWARD, *New Energy Technology – some Facts and Assessments* (London: MIT Press, 1971).

Index

Page numbers with the following letters added refer to:
b – select bibliography
n – notes
t – tables